GRADE 7

STP MATHEMATICS
for Jamaica

F W Ali   S Chandler   A Shepherd

L Bostock   C E Layne   E Smith

Nelson Thornes

Published in 2011 by:
Nelson Thornes Ltd
Delta Place
27 Bath Road
CHELTENHAM
GL53 7TH
United Kingdom

11 12 13 14 15 / 10 9 8 7 6 5 4 3 2 1

A catalogue for this book is available from the British Library

ISBN 978 1 4085 0912 8

Cover photos: Photolibrary

The authors and publishers are grateful to Janice Steele for her invaluable help with this book
Illustrations by Peters and Zabransky, Rupert Besley, Steve Ballinger, A & R Nelson, Guyana, Linda Jeffrey, Mike Bastin, OKS Group

Page make-up by The OKS Group

Printed and bound in Spain by GraphyCems

Contents

Introduction   vii

Preface   ix

1 Addition and subtraction of whole numbers   **1**

Place value   1
Ordering numbers   2
Continuous addition   4
Shortcuts   4
Addition of whole numbers   6
Shortcuts   6
Subtraction of whole numbers   9
Shortcuts   9
Mixed addition and subtraction   12
Approximation   14

2 Multiplication and division of whole numbers   **18**

Multiplication of whole numbers   18
Shortcuts   19
Multiplication by 10, 100, 1000, ...   20
Long multiplication   21
Shortcuts   22
Using a calculator for long multiplication   22
Division of whole numbers   24
Division by 10, 100, 1000, ...   26
Long division   27
Mixed operations of $+, -, \times, \div$ 28
Using brackets and order of operations   29
Number patterns   33
Types of number   36
Mixed exercises   38

3 Sets   **40**

Set notation   41
Describing members   41
The symbol $\in$ 43
The symbol $\notin$ 43
Finite and infinite sets   45
Equal sets   45
Empty set   46
Subsets   47
Universal set   47
Venn diagrams   49
Union of two sets   50
Intersection of sets   53

4 Factors and multiples   **56**

Factors   57
Multiples   57
Prime numbers   58
Divisibility tests   58
Highest Common Factor (HCF)   60
Lowest Common Multiple (LCM)   61
Shortcuts   61
Problems involving HCFs and LCMs   62

5 Fractions: addition and subtraction   **65**

The meaning of fractions   66
One quantity as a fraction of another   68
Equivalent fractions   69
Ordering fractions   72
Simplifying fractions   74
Shortcuts   74
Adding fractions   75
Fractions with different denominators   76
Subtracting fractions   78
Adding and subtracting fractions   79

Contents

Problems | 81
Mixed numbers and improper fractions | 82
Fractions as a result of division | 84
Adding mixed numbers | 84
Subtracting mixed numbers | 86
Mixed exercises | 87

6 Fractions: multiplication and division | 90

Multiplying fractions | 90
Simplifying | 91
Multiplying mixed numbers | 92
Whole numbers as fractions | 93
Fractions of quantities | 94
Dividing by fractions | 95
Dividing by whole numbers and mixed numbers | 96
Mixed multiplication and division | 97
Mixed operations | 98
Problems | 100
Mixed exercises | 101

7 Decimals | 104

The meaning of decimals | 104
Ordering decimals | 105
Changing decimals to fractions | 106
Changing fractions to decimals | 107
Addition of decimals | 107
Subtraction of decimals | 109
Multiplication by 10, 100, 1000, ... | 112
Division by 10, 100, 1000, ... | 113
Mixed multiplication and division | 114
Division by whole numbers | 115
Long division | 117
Changing fractions to decimals (exact values) | 117
Standard decimals and fractions | 118
Long method of multiplication | 119

Short method of multiplication | 119
Multiplication of decimals | 121
Problems | 122
Mixed exercises | 123

Review Test 1: Chapters 1–7 | 126

8 Metric units | 128

Units of length | 129
Some uses of metric units of length | 129
Changing from large units to smaller units | 131
Units of mass | 132
Mixed units | 133
Changing from small units to larger units | 134
Adding and subtracting metric quantities | 135
Accuracy of measurements | 137
Multiplying metric units | 138
Problems | 139
Time | 140
Temperature | 144
Mixed problems | 146
Mixed exercises | 148

9 Imperial units | 150

Units of length | 150
Units of mass | 152
Rough equivalence between metric and imperial units | 153

10 Introducing geometry | 156

Basic concepts | 157
Polygons | 157
Fractions of a revolution | 159
Bearings | 160

Angles 160
Right angles 161
Acute, obtuse and reflex angles 162
Degrees 163
Using a protractor to measure angles 164
Mixed questions 168
Drawing angles using a protractor 170
Angles on a straight line 171
Mixed exercises 174

11 Symmetry 177

Line symmetry 177
Congruency 178
Two axes of symmetry 181
Three or more axes of symmetry 183
Rotational symmetry (s-symmetry) 185

12 Introducing percentages 187

Expressing percentages as fractions 187
Expressing fractions as percentages 189
Problems 190
Expressing one quantity as a percentage of another 192
Finding a percentage of a quantity 193
Problems 194
Mixed exercises 196

13 Money matters 198

Money 198
Money units 199
Best buys 201
Profit and loss 203
Percentage profit and loss 204
Mixed exercises 206

14 Coordinates 208

Plotting points using positive coordinates 208

Quadrilaterals 213
Properties of the sides and angles of the special quadrilaterals 215

Review Test 2: Chapters 8–14 217

15 Area 220

Perimeter 221
Counting squares 224
Units of area 226
Compound figures 228
Problems 230
Changing units of area 233
Mixed problems 235

16 Solids 237

Solids 238
Volume and capacity 240
Nets 240
Practical work 244

17 Reflections 246

Reflections 246
Invariant points 249
Finding the mirror line 249
Drawing the mirror line 251
Strip patterns 254
Textile patterns 256

18 Introducing algebra 259

The idea of equations 259
Solving equations 261
Multiples of x 263
Mixed operations 264
Two operations 265
Problems 266
Simplifying expressions 268
Unlike terms 268

Contents

Equations with letter terms
on both sides   269
Equations containing like terms   270
Mixed exercises   272

19 More algebra   274

Brackets   274
Equations containing brackets   275
Hindu problem-solving   276
Problems to be solved by forming
equations   277
Formulae   279
Substituting numerical values
into a formula   282
Problems   283
Mixed exercises   284

20 Statistics   287

Frequency tables   287
Bar charts   290
Pie charts   294
Interpreting pie charts   297

Pictographs   298
Drawing pictographs   299

21 Probability   302

Outcomes of experiments   303
Probability   304
Experiments where an event
can happen more than once   305
Certainty and impossibility   308
Probability that an event
does not happen   309
Expectation   311

Review Test 3:
Chapters 15–21   316

Review Test 4:
Chapters 1–21   320

Answers   324

Index   343

Introduction

Why study mathematics?

Mathematics is found in:

(1) Your home
Mathematics is used in home building.
Frames are strengthened by triangle-bracing.
Geometric shapes are used to beautify
houses.

(2) Your diet
A good diet contains proper amounts
of basic food nutrients. Grams and
percentages of daily amounts of nutrients
are written on the packaging of foods.
Persons on special diets use metric scales to
measure their intake.

(3) Your career
The career you choose may require
mathematics. Some are Accountants,
Architects, Bank workers, Dieticians,
Draftsmen, Engineers, Electricians,
Carpenters, Surveyors and others.

(4) Your sports
We use mathematics to calculate batting and
other averages.
To find the chance of a victory we use probability.

THESE ARE BUT A FEW USES OF MATHEMATICS

This book attempts to satisfy your needs as you begin your study of mathematics
in the high school. We are very conscious of the need for success together with
the enjoyment everyone finds in getting things right. With this in mind we have
divided most of the exercises into three types of question:

The first type, identified by plain numbers, e.g. 12, helps you to see if you
understand the work. These questions are considered necessary for every
chapter you attempt.

The second type, identified by a single underline, e.g. 12, are extra, but not
harder, questions for quicker workers, for extra practice or for later revision.

The third type, identified by a double underline, e.g. <u><u>12</u></u>, are for those of you who manage Type 1 questions fairly easily and therefore need to attempt questions that are a little harder.

Most chapters end with 'mixed exercises'. These will help you revise what you have done, either when you have finished the chapter or at a later date.

All of you should be able to use a calculator accurately by the time you leave school. It is wise, in your first and second years, to use it mainly to check your answers, unless you have great difficulty with 'tables'. Whether you use the calculator or do the working yourself, always estimate your answer and always ask yourself the question, 'Is my answer a sensible one?'

Preface

To the teacher

The general aims of the series are:

(1) to help students to
- attain solid mathematical skills
- connect mathematics to their everyday lives and understand its role in the development of our contemporary society
- see the importance of thinking skills in everyday problems
- discover the fun of doing mathematics and reinforce their positive attitudes to it.

(2) to encourage teachers to include historical information about mathematics in their programme.

In writing this three-book series the authors attempted to present topics in such a way that students will understand the connections in mathematics, and to be encouraged to see and use mathematics as a means to help make sense in the real world.

Topics from the history of mathematics have been incorporated to ensure that mathematics is not dissociated from its past. This should lead to an increase in the levels of enthusiasm, interest and fascination for mathematics, and should also enrich the teaching of it.

Careful grading of exercises makes the books approachable.

Some suggestions:

(1) Before each lesson give a brief outline of the topic to be covered in the lesson. As examples are given refer back to the outline to show how the example fits into it.

(2) List terms on the chalkboard that you consider new to the students. Solicit additional words from the class and encourage students to read from the text and make their own vocabulary.
Remember that mathematics is a foreign language. The ability to communicate mathematically must involve the careful use of the correct terminology.

(3) When possible have students construct alternative ways to phrase questions. This ties in with seeing mathematics as a language. Students, especially in the junior classes, tend to concentrate on the numerical or 'maths' part of the question and pay little attention to the instructions that give information that is required to solve the problem.

(4) When solving problems have students identify their own problem-solving strategies and listen to others. This practice should create an atmosphere of discussion in the class, centred around different approaches to the same problem.

As the students try to solve problems on their own they will make mistakes. This is healthy, as this was the experience of the inventors of mathematics: they tried, guessed, made many mistakes and worked for hours, days and sometimes years before reaching a solution.

There are enough problems in the exercises to allow the students to try and try again. The excitement, disappointment and struggle with a problem until a solution is found provide a healthy classroom atmosphere.

 Investigation

'Investigation' is included in these books. This is in keeping with the requirements of the latest CSEC syllabus.

Investigation is used to provide students with the opportunity to explore hands-on and minds-on mathematics. At the same time teachers are presented with open-ended explorations to enhance their mathematical instruction.

It is expected that the tasks will
• invite problem-solving and reasoning
• require communication skills
• connect various mathematical concepts and principles.

It should also involve processes including collecting data, collaborating with peers, and using multiple strategies to reach conclusions.

A mathematical investigation should also meet the following requirements:

It should be
• multidimensional in content
• open-ended, with several acceptable solutions
• centred on a theme or event
• embedded in a focus question.

To the student

These books are written for you. As you study:
• Try to break up the material in a chapter into manageable bits.
• Always have paper and pencil when you study mathematics.
• When you meet a new word write it down together with its meaning.
• Read your questions carefully and rephrase them in your own words.

- The information that you need to solve your problem is given in the wording of the problem, not only in the number part.
- Your success in mathematics may be achieved through practice.
- You are therefore advised to try to solve as many problems as you can.
- Always try more problems than those set by your teacher for homework.

Remember that the greatest cricketer or netball player became great by practising for many hours.

We have provided enough problems in the books to allow you to practise. Above all, don't be afraid to make mistakes as you are learning. The greatest mathematicians all made many mistakes as they tried to solve problems.

You are now on your way to success in mathematics – GOOD LUCK!

1 Addition and subtraction of whole numbers

At the end of this chapter you should be able to...

1 Recognise the place value of a digit.

2 Order numbers.

3 Add and subtract whole numbers.

4 Approximate a given whole number to the nearest ten, hundred or thousand.

5 Solve problems involving addition and subtraction of whole numbers.

6 Use a calculator to add and subtract whole numbers after finding approximate answers.

You need to know...

✔ pairs of numbers that add up to 10

✔ pairs of numbers that add up to 100

✔ the sum of any two numbers less than 10

Key words

difference, digit, hundreds, numeral, place value, tens, units

Place value

A number written in symbols is called a numeral.

The symbols we use are the digits 0, 1, 2, 3, 4, 5, 6, 7, 8, 9.

You can write the number one thousand four hundred and twenty-seven as 1427.

The number two thousand seven hundred and forty-one is 2741.

The same digits are used but they are in different places.

In 1427, the digit 2 means 2 tens.

In 2741, the digit 2 means 2 thousands.

You can write a number under place headings,

	thousands	hundreds	tens	units
e.g. 1427 can be written	1	4	2	7
856 can be written		8	5	6
7502 can be written	7	5	0	2
and 3010 can be written	3	0	1	0

Ordering numbers

Ordering numbers means placing them in order of size.
Ascending order means the smallest number first, then the next smallest, and so on.
Descending order has the largest number first, the next largest, and so on.

Exercise 1a

1   Write these numbers using digits.
   a sixty-three   **b** forty-nine
   c seven hundred and seven   **d** three hundred and twenty-seven

2   Write these numbers using digits.
   a eight hundred and nineteen   **b** eight thousand and eight
   c six thousand and sixty-seven   **d** fifteen thousand, two hundred and thirty-four

3   Write in words:
   a 56   **b** 79   **c** 409   **d** 187   **e** 734

4   Write in words:
   a 330   **b** 426   **c** 9488   **d** 6593   **e** 7065

5   Look at the number 96 538.
   Write down the digit that gives
   a the number of hundreds   **b** the number of thousands.

6   Look at the number 70 869.
   Write down the digit that gives
   a the number of units   **b** the number of thousands
   c the number of hundreds.

7   Write these numbers in ascending order.
   a 55, 43, 63, 57
   b 83, 31, 49, 27
   c 308, 77, 293, 104

The number with the smallest number of 10s is 43 so 43 is the smallest number. There are two numbers in the 50s. 55 comes before 57 so 55 is the next smallest number after 43. 63 has the largest number of 10s so 63 is the largest number of the four.

8 Write these numbers in ascending order.

    a 506, 605, 650, 560   **b** 845, 8876, 98, 1088   **c** 2303, 3302, 3032, 2033

9 Write down the place value of the 4 in these numbers.

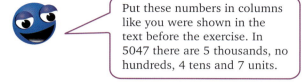

Put these numbers in columns like you were shown in the text before the exercise. In 5047 there are 5 thousands, no hundreds, 4 tens and 7 units.

    a 5047   **b** 6403

    c 3304   **d** 4056

    e 48 976

10 Write down the place value of the 6 in these numbers.

    a 3607   **b** 9056   **c** 6883   **d** 62 854

11 Use all three of these cards to make

    a the largest number possible

    b the smallest number possible.

12 Use all three of these cards to make

    a the largest number possible

    b the smallest number possible.

054 is not an acceptable number.

13 Use the digits 3, 7, 9, 2 once each to make the smallest possible number.

14 Use the digits 5, 6, 2, 3 once each to make the largest possible number.

15 Use the digits 7, 0, 5, 1 once each to make the smallest possible number bigger than five thousand.

16 Use the digits 4, 6, 8, 5 once each to make the largest possible number smaller than six thousand.

17 Look at these numbers.

Write all the numbers using digits, then look at the thousands, then the hundreds, and so on.

    two thousand four hundred, 2040

    two thousand and forty-four, 2440

    two thousand two hundred and forty

    a Which is the largest?   **b** Which is the smallest?

18 Write down as many different numbers as you can using each of the digits 3, 4, 5 once in each number. Put these numbers in descending order.

19 Write down as many different numbers as you can using each of the digits 2, 6, 7 once in each number. Put these numbers in order with the smallest first.

We use whole numbers all the time in everyday life and it is important that we should be able to add them and subtract them accurately in our heads. This comes with practice.

Continuous addition

To add a line of numbers, start at the left-hand side:

Working in your head

$6+4+3+8 = 21$ add the first two numbers (10)

then add on the next number (13)

then add on the next number (21).

Check your answer by starting at the other end.

To add a column of numbers, start at the bottom and *working in your head* add up the column:

$$
\begin{array}{r}
8 \\
7 \\
2 \\
+5 \\
\hline
22 \\
\hline
\end{array}
$$

$(5+2 = 7, 7+7 = 14, 14+8 = 22)$

Check your answer by starting at the top and adding down the column.

Shortcuts

You can add numbers in any order so look for pairs of numbers that add up to ten. Then add up other pairs.

For example, $8+7+2+5 = 10+12 = 22$

Exercise 1b

Find the value of:

1 $2+3+1+4$

2 $1+5+2+3$

3 $5+2+6+1$

4 $3+4+2+6$

5 $5+6+4+2$

6 $8+2+9+5$

<u>7</u> $7+3+8+6$

<u>8</u> $5+4+9+1$

<u>9</u> $7+3+2+8$

<u>10</u> $6+7+5+9$

<u>11</u> $2+5+4+1+3$

<u>12</u> $4+8+2+1+2$

<u>13</u> $6+7+3+5+6$

<u>14</u> $4+9+2+8+4$

<u>15</u> $7+3+9+6+8$

<u>16</u> $3+2+3+4+1+5$

<u>17</u> $4+2+5+6+1+7$

<u>18</u> $8+3+9+2+7+3$

<u>19</u> $6+9+4+8+7+5$

<u>20</u> $4+7+8+6+5+2$

21
$$
\begin{array}{r}
3 \\
7 \\
8 \\
+6 \\
\hline
\end{array}
$$

22
$$
\begin{array}{r}
1 \\
9 \\
5 \\
+2 \\
\hline
\end{array}
$$

23
$$
\begin{array}{r}
4 \\
6 \\
7 \\
+3 \\
\hline
\end{array}
$$

24
$$
\begin{array}{r}
9 \\
7 \\
9 \\
+8 \\
\hline
\end{array}
$$

25
$$
\begin{array}{r}
8 \\
7 \\
6 \\
+9 \\
\hline
\end{array}
$$

26	3	**27**	4	**28**	6	**29**	7	**30**	6
	4		2		5		8		7
	5		3		3		2		3
	1		9		1		1		9
	+8		+3		+4		+8		+7
	─		─		─		─		─

31	3	**32**	5	**33**	8	**34**	2	**35**	4
	5		7		7		9		8
	2		3		9		5		2
	9		5		2		8		9
	1		4		8		7		9
	+6		+2		+6		+6		+7
	─		─		─		─		─

? Puzzle

Each disc shows three digits. Remove one digit from one of the discs and place it on another disc so that the digits on each of the three discs have the same total.

? Puzzle

Copy this diagram onto a sheet of paper.

Cut it into three pieces and fit them together to form a magic square.

 In a magic square, the numbers in each row and in each column and in each diagonal have the same total.

```
        7       5
   9    2       3   4
        6   1   8
```

! Investigation

Sonia makes up a pattern starting with 3 and 4. To get the next number in the pattern she adds the previous two numbers together. If the answer is more than 10 she writes down only the number units.

Her pattern is 3, 4, 7, 1, 8, 9, 7...

Write down the next ten numbers in this pattern. Does the pattern repeat itself? If so, how many numbers are there before it starts to repeat?

Now start with 4 and 3 and see what happens. (You need to keep going for a long time!)

Investigate some other pairs of numbers.

Addition of whole numbers

To add a column of numbers, start with the units:

In the *units* column, $2+1+3=6$ so write 6 in the units column.

```
   8 3
   2 9 1
 + 7 0 2
 ───────
 1 0 7 6
     ₁
```

In the *tens* column, $0+9+8=17$ tens which is 7 tens and 1 hundred. Write 7 in the tens column and carry the 1 hundred to the hundreds column to be added to what is there already.

In the *hundreds* column, $1+7+2=10$ hundreds which is 0 hundreds and 1 thousand.

Shortcuts

You can add the tens and units separately.

For example, $23+48+16=(20+40+10)+(3+8+6)$

$$= \qquad 70 \qquad + \quad 17 \quad = 87$$

You can do the same with larger numbers.

For example, $264+38+509=(200+500)+(60+30+0)+(4+8+9)$

$$= \qquad 700 \quad + \qquad 90 \quad + \quad 21$$

$$= 790+21 = 811$$

Exercise 1c

Find the value of the following sums:

	1		2		3		4		5	
	28		35		22		103		56
	+51		+62		+43		+205		+203

	6		7		8		9		10	
	101		223		492		259		351
	25		317		812		28		1026
	+273		+342		+735		+704		+915

	11		12		13		14		15	
	87		93		3021		9217		6943
	102		251		84		824		278
	56		179		926		3216		5419
	+304		+1312		+5041		+8572		+3604

16	$28 + 72 + 12$	**31**	$694 + 706 + 293$
17	$56 + 10 + 92$	**32**	$325 + 576 + 481$
18	$83 + 107 + 52$	**33**	$253 + 431 + 1212$
19	$256 + 139 + 402$	**34**	$821 + 903 + 3506$
20	$1026 + 398 + 542$	**35**	$727 + 652 + 2716$
21	$24 + 83 + 76$	**36**	$92 + 56 + 109 + 324$
22	$92 + 58 + 27$	**37**	$103 + 72 + 58 + 276$
23	$52 + 112 + 38$	**38**	$329 + 26 + 73 + 429$
24	$207 + 394 + 651$	**39**	$256 + 82 + 712 + 37$
25	$943 + 856 + 984$	**40**	$325 + 293 + 502 + 712$
26	$826 + 907 + 329$	**41**	$624 + 1315 + 437 + 516$
27	$562 + 497 + 208$	**42**	$2514 + 397 + 3617 + 251$
28	$599 + 107 + 2058$	**43**	$752 + 593 + 644 + 237$
29	$642 + 321 + 4973$	**44**	$2516 + 374 + 527 + 152$
30	$555 + 921 + 6049$	**45**	$879 + 4658 + 5743 + 652$

Numbers are easier to add when they are written in a column.
Make sure that the units are lined up in one column, the tens in the next column, and so on.

Exercise 1d

Find the total cost of some cookies priced at one hundred and six dollars and a book priced at one thousand and forty-three dollars.

To find the total cost, you need to add the price of the book to the price of the cookies.

First you have to write the prices in digits: one hundred and six means 1 hundred, 0 tens and 6, i.e. 106.
One thousand and forty-three means 1 thousand, 0 hundreds, 4 tens and 3, i.e. 1043.

Total cost is $106 + $1043

$$\begin{array}{r} 106 \\ +1043 \\ \hline 1149 \end{array}$$

i.e. $1149

1 Find the total cost of a tin of baked beans at $58, a cake at $36 and a can of cola at $65.

2 In the local corner shop I bought a comic costing $65, a pencil costing $45 and a packet of sweets costing $50. How much did I spend?

3   There are three classes in the first year of a school. One class has 29 children in it, another class has 31 children in it and the third class has 28 children in it. How many children are there in the first year of the school?

4   Find the total cost of a washing machine at $21 420, a cooker at $17 990 and a fridge at $13 580.

5   Write the following numbers in digits:
    a   two hundred and sixty-one   **b**   three hundred and two
    c   three thousand and fifty-six   **d**   thirteen hundred.

6   Write the following numbers in words:
    a   324   **b**   5208   **c**   150   **d**   1500.

7   Add four hundred and fifteen, one hundred and sixty-eight and two hundred and four.

8   I have three pieces of string. One piece is 27 cm long, another piece is 34 cm long and the third piece is 16 cm long. What is the total length of string that I have?

9   Find the total cost of a calculator at $1200, a pencil set at $500 and a cassette at $3600.

10   When John went to school this morning it took him 4 minutes to walk to the bus stop. He had to wait 12 minutes for the bus and the bus journey took 26 minutes. He then had an 8 minute walk to his school. How long did it take John to get to school?

11   Find the sum of one thousand and fifty, four hundred and seven and three thousand five hundred.

12   A boy decided to save some money by an unusual method. He put $1 in his money box the first week, $2 in the second week, $4 in the third week, $8 in the fourth week, and so on. He gave up after 10 weeks. Write down how much he put in his money box each week and add it up to find the total that he had saved. Why do you think he gave up?

! Investigation

A number that reads the same forwards and backwards, e.g. 14 241, is called a palindrome.

There is conjecture (meaning, it has not been proved) that if we take *any* number, reverse the digits and add the numbers together, then do the same with the result, and so on, we will end up with a palindrome.
For example, start with 251 $251 + 152 = 403$
                                                     $403 + 304 = 707$ which is a palindrome.

1 Can you find some two-digit numbers for which the palindrome appears after the first sum?

2 If two palindromes are added together, is the result always a palindrome? Give a reason for your answer.

Subtraction of whole numbers

Exercise 1e

Do the following subtractions in your head:

1	15 − 4	2	19 − 7	3	18 − 4	4	12 − 7	5	15 − 8
6	20 − 8	9	14 − 6	12	12 − 9	15	19 − 9	18	15 − 9
7	18 − 3	10	10 − 4	13	17 − 6	16	11 − 7	19	20 − 6
8	17 − 8	11	15 − 2	14	16 − 8	17	13 − 8	20	15 − 7

You will probably have your own method for subtraction.
Use it if you understand it.
Here is one method:

$$\begin{array}{r} {}^{0\;14\;1}\!\!\!\!\!15\,0\,8 \\ -\;7\,2\,1 \\ \hline 7\,8\,7 \end{array}$$

Start with the units column, then do the tens column, and so on.
If you cannot do the subtraction, take one from the top number in the next column; this is worth ten in the column on its right.

Shortcuts

You can work out a subtraction using addition. You do this by counting on.
For example, to find $109 - 64$, count on from 64 until you get to 109:

$$64 \rightarrow 70 \rightarrow 100 \rightarrow 109$$
$$6 + 30 + 9 = 45$$

So $109 - 64 = 45$.

Exercise 1f

Find:

1 $526 - 315$

2 $754 - 203$

3 $821 - 415$

4 $526 - 308$

5 $564 - 491$

6 $395 - 254$

7 $708 - 302$

8 $495 - 369$

It is easier to subtract numbers when they are written in columns. Remember to place units under one another, tens under one another, and so on.

9	283 − 157	**19**	718 − 439	**29**	4627 − 3924
10	638 − 452	**20**	308 − 159	**30**	1203 − 527
11	814 − 344	**21**	507 − 499	**31**	4906 − 829
12	592 − 238	**22**	3451 − 623	**32**	1516 − 468
13	578 − 291	**23**	5267 − 444	**33**	3506 − 3429
14	635 − 457	**24**	7374 − 759	**34**	7016 − 6824
15	602 − 415	**25**	1027 − 452	**35**	9342 − 5147
16	704 − 568	**26**	3927 − 583	**36**	6309 − 4665
17	1237 − 524	**27**	1922 − 398		
18	823 − 568	**28**	2704 − 2515		

Exercise 1g

Cheryl had 49 marbles in her bag. Darren had 86 marbles in his bag. How many more marbles did Darren have than Cheryl had?

'How many more' means the number that Darren has over and above 49, i.e. you have to take 49 from 86.

86 − 49 = 37

So, Darren had 37 more marbles than Cheryl.

To find 86 − 49 in your head, take 50 off 86 then add 1.

1 The club dues for last week were $970. I paid with a $5000 note. How much change should I have?

Read the question carefully to make sure that you understand what you are being asked to do. Read it as many times as you need to.

2 In a school there are 856 children. There are 392 girls. How many boys are there?

3 Find the difference between 378 and 293.

4 Take two hundred and fifty-one away from three hundred and forty.

5 A shop starts with 750 cans of cola and sells 463. How many cans are left?

6 Subtract two thousand and sixty-five from eight thousand, five hundred and forty-eight.

7 Find the difference between 182 and 395.

8 The road I live on has 97 houses. The road my friend lives on has 49 houses. How many more houses are there on my road than on my friend's road?

9 The highest peak of the Blue Mountains in Jamaica is 2220 m. Mount Everest is 8843 m high. How much higher than the highest Blue Mountain peak is Mount Everest?

10 My brother is 123 cm tall and I am 142 cm tall. What is the difference between our heights?

! Investigation

Write down any three-digit number, e.g. 287.

Arrange the digits in order of size – once with the largest digit first and once with the smallest digit first, i.e. 872, 278.

Now find the difference between these two numbers (take the smaller number from the larger number),

i.e. $872 - 278 = 594$.

Repeat the process for your answer,

i.e. $954 - 459 = 495$.

Repeating the process again,

i.e. $954 - 459 = 495$.

These rules give us a chain of numbers.

In this example we have $287 \rightarrow 594 \rightarrow 495 \rightarrow 495$.

Form a similar chain for a three-digit number of your own choice.
Try a few more. What do you notice?

Exercise 1h

Find the missing digit – it is marked with □:

1 $27 + 38 = \square 5$

2 $34 + 5\square = 89$

3 $5\square - 25 = 32$

4 $6\square - 48 = 16$

5 $128 + \square 59 = 1087$

6 $5\square + 29 = 83$

7 $\square 4 + 57 = 81$

8 $\square 3 - 47 = 26$

9 $25 - 1\square = 6$

10 $1\square 7 + 239 = 416$

Mixed addition and subtraction

It is the sign *in front* of a number that tells you what to do with that number.
For example, $128-56+92$ means '128 take away 56 and add on 92'. This
can be done in any order with addition and subtraction so we could add on
92 and then take away 56.

$$128-56+92 = 220-56$$
$$= 164$$

Exercise 1i

Find $138+76-94$

$$138+76-94 = 214-94$$
$$= 120$$

$$\begin{array}{rr} 138 & 214 \\ +\ 76 & -\ 94 \\ \hline 214 & 120 \end{array}$$

Find $56-72+39-14$

$$56-72+39-14 = 56+39-72-14$$
$$= 95-72-14$$
$$= 23-14$$
$$= 9$$

$$\begin{array}{rr} 56 & 95 \\ +39 & -72 \\ \hline 95 & 23 \end{array}$$

Find:

1	$25-6+7-9$	**11**	$17-9+11-19$	**21**	$213-307+198-31$
2	$14+2-8-3$	**12**	$36-24+62-49$	**22**	$29+108-210+93$
3	$7-4+5-6$	**13**	$51-27-38+14$	**23**	$493-1000+751-140$
4	$19+2-4+3$	**14**	$43-29+37+16$	**24**	$36+52-73+29-37$
5	$23-2+4+5$	**15**	$124+51-78-14$	**25**	$78-43+15-39+18$
6	$46-12+3-9$	**16**	$91-50+36-27$	**26**	$612-318+219+84$
7	$27+6-11-9$	**17**	$105+23-78-50$	**27**	$95-161+75+10$
8	$2+13-7+3-8$	**18**	$73-42-19+27$	**28**	$952-1010-251+438$
9	$7-6+9-1-3$	**19**	$215-181+36-70$	**29**	$278+394-506+84$
10	$17+4-9-3-5$	**20**	$361-200+15-81$	**30**	$107-1127+854+231$

? Puzzle

Solve the following cross-number puzzle.

Across 1. $73 - 31$
        4. $249 + 167$
        6. $700 - 565$
        7. $231 - 158$
Down 2. $77 + 166$
        3. $222 - 136$
        5. $78 + 79$
        6. $52 + 106 - 139$

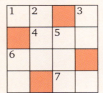

Exercise 1j

A shop had 500 bottles of water in stock at closing time on Monday. On Tuesday 31 bottles were sold, on Wednesday 84 bottles were sold and on Thursday 69 bottles were sold. How many bottles were left in stock by closing time on Thursday?

The number of bottles left is the difference between 500 and the total sold on Tuesday, Wednesday and Thursday.
First find this total: i.e. $31 + 84 + 69$ which is 184.
The number of bottles left is $500 - 184 = 316$.

1 A boy buys a comic costing $5 and a pencil costing $2. He pays with a $50 note. How much change does he get?

These questions involve more than one operation; do not try to do them all at once.

2 Find the sum of eighty-six and fifty-four and then take away sixty-eight.

3 I have a piece of string 200 cm long. I cut off two pieces, one of length 86 cm and one of length 34 cm. How long is the piece of string that I have left?

4 On Monday 1000 fish fingers were cooked in the school kitchen. At the first dinner sitting 384 fish fingers were served. At the second sitting 298 fish fingers were served. How many were left?

5 Find the difference between one hundred and ninety and eighty-three. Then add on thirty-seven.

6 A minimart has 38 kg of carrots when it opens on Monday morning. During the day the shop gets a delivery of 60 kg of carrots and sells

29 kg of carrots. How many kilograms of carrots are left when it closes on Monday evening?

7 A boy has 30 marbles in his pocket when he goes to school on Monday morning. At morning break he wins 6 marbles. At lunchtime he loses 15 marbles. After school he loses 4 marbles. How many marbles does he now have?

8 What is three hundred and twenty-seven plus two hundred and six minus four hundred and eighty-eight?

9 Sarah gets $50 pocket money on Saturday. On Monday she spends $34. On Tuesday she is given $20 for doing a special job at home. On Thursday she spends $27. How much money has she got left?

10 Yesterday Kevin reached his 12th birthday. Next year he will be 14. What is today's date, and when is Kevin's birthday?

! Investigation

Follow these instructions:

Step 1 Write down any three-digit number in which the number of hundreds differs by at least two from the number of units, e.g. 419 or 236 or 973 but not 707 or 514.

Step 2 Now write down the digits in reverse order, e.g. 419 becomes 914.

Step 3 Subtract the smaller number from the larger number, e.g. 914 − 419 = 495.

Step 4 Add this number to its reverse, i.e. add 495 to 594.

The result in this case is 1089.

Now try these four steps with any number of your choice. Repeat the instructions six times for six different numbers, but do remember that the number of hundreds must differ by at least two from the number of units.

What do you notice?

Approximation

Calculators are very useful and can save a lot of time. Calculators do not make mistakes but *we* sometimes do when we use them. So it is important to know roughly if the answer we get from a calculator is right. By simplifying the numbers involved we can get a rough answer in our heads.

One way to simplify numbers is to make them into the nearest number of tens. For example

127 is roughly 13 tens, or 130

and

123 is roughly 12 tens, or 120

We say that 127 is *rounded up* to 130 and 123 is *rounded down* to 120.
In mathematics we say that 127 is approximately equal to 13 tens.

We use the symbol ≈ to mean 'is approximately equal to'. We would write

$127 \approx 13$ tens

$123 \approx 12$ tens

When a number is half way between tens we always round up. We say

$125 \approx 13$ tens

Exercise 1k

Write each of the following numbers as an approximate number of tens:

$56 \approx 6$ tens

1	84	**3**	46	**5**	8	**7**	228	**9**	73
2	151	**4**	632	**6**	37	**8**	155	**10**	4

Write each of the following numbers as an approximate number of hundreds:

$1278 \approx 13$ hundreds

11	830	**13**	780	**15**	1350	**17**	1560	**19**	972
12	256	**14**	1221	**16**	450	**18**	3780	**20**	1965

By writing each number correct to the nearest number of tens find an approximate answer for:

$196 + 58 - 84$

$196 \approx 20$ tens

$58 \approx 6$ tens

$84 \approx 8$ tens

Therefore $\qquad 196 + 58 - 84 \approx 20$ tens $+ 6$ tens $- 8$ tens

≈ 18 tens $= 180$

21	$344 - 87$	**31**	$83 + 27 - 52$	**41**	$127 + 56 + 82 + 95$
22	$95 - 39$	**32**	$76 - 31 - 29$	**42**	$73 + 21 + 37 + 46 + 29$
23	$258 - 49$	**33**	$137 - 56 + 82$	**43**	$33 + 18 + 27 + 96 + 53$
24	$472 + 35$	**34**	$241 + 37 - 124$	**44**	$13 + 29 + 83 + 121 + 5$
25	$153 + 181$	**35**	$295 + 304 - 451$	**45**	$41 + 82 + 96 + 73 + 36$
26	$89 - 51$	**36**	$49 - 25 + 18$	**46**	$83 + 64 + 95 + 51$
27	$258 + 108$	**37**	$68 + 143 + 73$	**47**	$63 + 29 + 40 + 37 + 81$
28	$391 - 127$	**38**	$153 + 19 + 57$	**48**	$108 + 16 + 29 + 53 + 85$
29	$275 - 99$	**39**	$369 - 92 + 85$	**49**	$17 + 23 + 46 + 9 + 75$
30	$832 - 55$	**40**	$250 + 31 - 121$	**50**	$103 + 125 + 76 + 41 + 8$

Now use your calculator to find the exact answers to numbers **21** to **50**. Remember to look at your rough answer to check that your calculator answer is probably right.

Did you know?

The Romans did not use symbols for numbers, but used letters of the alphabet. For example, the Romans used X for ten, V for five – XV means 'ten and five', i.e. 15.

The Roman way of writing numbers is still used today. (When you write your CXC examinations, your grades are written using Roman numerals. If you study hard and do very well you will get Grade I.)

The numbers one to six are written I, II, III, IV, V, VI.

1 Use reference material to find out what the following Roman numbers are: XIV, CLX, MLII.

2 Write the following numbers in Roman numerals: 25, 152, 1854, 2006.

3 Find LXII – XXIV and write your answer in Roman numerals.

4 In Roman numerals, the year 2011 is written MMXI.
   Which year, written in Roman numerals, uses the most letters?
   Hint: it is in the 19th century.

In this chapter you have seen that...

✔ numbers are easier to add or subtract when they are written in columns (unless you can do the calculation mentally)

✔ you can do a rough check on your calculations by rounding the numbers to the nearest 10, 100 or 1000

✔ to solve word problems, you need to read the question carefully to make sure that you understand what you have been asked to do

✔ the sign in front of a number tells you whether to add or subtract that number.

2 Multiplication and division of whole numbers

At the end of this chapter you should be able to...

1 Multiply or divide one whole number by another.

2 Estimate the product of two whole numbers to the nearest ten, hundred or thousand.

3 Perform operations involving a combination of $+$, $-$, $\times$ and $\div$

4 Supply numbers to continue a given sequence.

5 Solve problems using brackets.

6 Identify square, rectangular and triangular numbers.

You need to know...

✔ the multiplication tables up to 10×10

✔ about place value

Key words

bracket, calculator, consecutive, estimate, even number, magic square, magic triangle, number pattern, odd number, operation, pyramid, rectangular number, remainder, sequence, square number, triangular number.

Multiplication of whole numbers

The following exercise will help you to practise the multiplication facts.

For example 69×4 can be found by adding

$$69 + 69 + 69 + 69$$

but it is quicker to use the multiplication facts.

Break 69 into $60 + 9$.

Then $69 \times 4 = (60 \times 4) + (9 \times 4)$
$= 240 + 36$
$= 276$

The traditional way of setting this out is:

$$\begin{array}{r} 69 \\ \times \quad 4 \\ \hline 276 \\ \hline \end{array}$$
3

Shortcuts

There are shortcuts for some numbers.
Multiplying by 2 is easy. You double the answer.

To multiply by 4, double then double again.

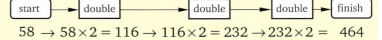

For example, to find 58×4:

$58 \ \rightarrow \ 58 \times 2 = 116 \ \rightarrow \ 116 \times 2 = \ 232$

To multiply by 8, double three times.

start → double → double → double → finish

For example, to find 58×8:

$58 \rightarrow 58 \times 2 = 116 \rightarrow 116 \times 2 = 232 \rightarrow 232 \times 2 = \ 464$

To multiply by 6, multiply by 3 then double.

start → ×3 → double → finish

For example, to find 58×6:

$58 \rightarrow 58 \times 3 = 174 \rightarrow 174 \times 2 = \ 348$

To multiply by 5, you can multiply by ten, then halve the result.

start → ×10 → halve → finish

For example, to find 58×5:

$58 \rightarrow 58 \times 10 = 580 \rightarrow 580 \div 2 = \ 290$

Exercise 2a

Find:

If you can do these in your head, write down the answer. You may find it easier to write the numbers in columns. Keep units under one another, tens under one another and so on.

For example
$$\begin{array}{r} 23 \\ \times \ 2 \\ \hline \end{array}$$

1	23×2	**4**	76×4	**7**	25×4	**10**	83×5
2	42×3	**5**	58×5	**8**	16×9	**11**	47×3
3	13×8	**6**	31×3	**9**	72×2	**12**	54×6

13	21×6	**23**	7×32	**33**	876×3		
14	84×7	**24**	9×27	**34**	312×7		
15	36×9	**25**	152×4	**35**	142×6		
16	73×4	**26**	307×8	**36**	513×5		
17	2×81	**27**	256×3	**37**	6×529	**43**	848×8
18	33×4	**28**	194×2	**38**	857×6	**44**	9×659
19	67×8	**29**	221×9	**39**	7×498	**45**	748×7
20	73×9	**30**	211×4	**40**	579×9	**46**	694×8
21	49×6	**31**	953×3	**41**	658×7	**47**	236×7
22	8×21	**32**	204×8	**42**	7×427	**48**	573×9

> Remember that you can change the order of the numbers.
> So 8×496 is the same as 496×8

⚠ Investigation

 1 *Do not use a calculator for parts **a** to **d**.*

    a Multiply 123 456 789 by 3 and then multiply the result by 9. What do you notice?

    b Repeat part **a** multiplying first by a different number less than 9.

    c Repeat part **a** again using a third number less than 9.

    d Is there a rule for predicting the answer when 123 456 789 is multiplied by one of the numbers 2, 3, 4, 5, 6, 7 or 8, and the result is multiplied by 9? If you find one, write it down and test it.

2 *Now try using your calculator*. What do you notice?

Multiplication by 10, 100, 1000, …

When 85 is multiplied by 10 the 5 units become 5 tens and the 8 tens become 8 hundreds. So

$$85 \times 10 = 850$$

When 85 is multiplied by 100 the 5 units become 5 hundreds and the 8 tens become 8 thousands. Thus

$$85 \times 100 = 8500$$

When 85 is multiplied by 20 this is the same as $85 \times 2 \times 10$. So

$$85 \times 20 = 85 \times 2 \times 10$$
$$= 170 \times 10$$
$$= 1700$$

In the same way

$$27 \times 4000 = 27 \times 4 \times 1000$$
$$= 108 \times 1000$$
$$= 108\,000$$

Exercise 2b

Find:

1	27×10	**13**	221×30	**25**	390×90
2	82×100	**14**	127×700	**26**	107×400
3	36×10	**15**	73×2000	**27**	240×80
4	108×10	**16**	39×900		
5	256×1000	**17**	157×60		
6	27×20	**18**	295×80		
7	82×300	**19**	88×70		
8	51×40	**20**	350×200		
9	39×200	**21**	609×80		
10	56×50	**22**	270×200	**28**	100×88
11	73×400	**23**	556×70	**29**	200×95
12	58×60	**24**	81×3000	**30**	856×70

 To multiply by 20, multiply by 2 first, then multiply your answer by 10.

 To multiply by 300, multiply by 3 first, then multiply your answer by 100.

Long multiplication

To multiply 84×26 we use the fact that

$$84 \times 26 = 84 \times 20 + 84 \times 6$$

This can be set out as

```
        84
      × 26
       504     (84 × 6)
     +1680     (84 × 20)
      2184
```

Shortcuts

This method involves only multiplying and dividing by 2 and adding up.
It works with any two numbers.

84	26
42	52
21	104*
10	208
5	416*
2	832
1	1664*

To find 84×26 start by writing the numbers next to each other.
Divide the left-hand number by 2 and multiply the right-hand number by 2.
Repeat this, ignoring any remainders, until you end up with 1 on the left.

Put an asterisk against numbers in the right-hand column that are next
to odd numbers.

Add the numbers with asterisks:

$104 + 416 + 1664 = 2184$

So $84 \times 26 = 2184$

Large numbers are easier to multiply together when written in columns. Remember to keep units lined up in a units column, tens in a tens column.

Exercise 2c

Find:

1	32×21	**11**	241×32	**21**	2004×43
2	43×13	**12**	153×262	**22**	584×97
3	86×15	**13**	433×921	**23**	187×906
4	27×21	**14**	1251×28	**24**	270×709
5	34×42	**15**	3421×33	**25**	3060×470
6	38×41	**16**	512×210	**26**	385×95
7	107×26	**17**	487×82	**27**	750×450
8	53×82	**18**	724×98	**28**	605×750
9	74×106	**19**	146×259	**29**	1008×908
10	36×89	**20**	805×703	**30**	1500×802

Using a calculator for long multiplication

Calculators save a lot of time when used for long multiplication. You do,
however, need to be able to estimate the size of the answer you expect as a
check on your use of the calculator.

One way to get a rough answer is to round off

a number between 10 and 100 to the nearest number of tens

a number between 100 and 1000 to the nearest number of hundreds

a number between 1000 and 10 000 to the nearest number of thousands

and so on.

For example $512 \times 78 \approx 500 \times 80 = 40\,000$

and $2752 \times 185 \approx 3000 \times 200 = 600\,000$

Exercise 2d

Estimate:

1	79×34	**6**	59×18	**11**	159×93
2	29×27	**7**	23×55	**12**	82×309
3	84×36	**8**	62×57	**13**	281×158
4	45×32	**9**	136×29	**14**	631×479
5	87×124	**10**	52×281	**15**	273×784

Estimate the answer and then use your calculator to work out the following:

2581×39

$$2581 \times 39 \approx 3000 \times 40 = 120\,000 \quad \text{(estimate)}$$

$$2581 \times 39 = 100\,659 \quad \text{(calculator)}$$

16	258×947	**24**	2501×12	**32**	47×853	**40**	68×529
17	29×384	**25**	87×76	**33**	94×552	**41**	37×634
18	182×56	**26**	52×821	**34**	18×47	**42**	541×428
19	37×925	**27**	89×483	**35**	62×98	**43**	798×583
20	782×24	**28**	481×97	**36**	463×87	**44**	694×7281
21	78×91	**29**	608×953	**37**	271×82	**45**	7215×48
22	625×14	**30**	4897×61	**38**	753×749		
23	33×982	**31**	69×78	**39**	492×47		

Exercise 2e

A library has 68 shelves. There are 73 books on each shelf. How many books are there altogether?

To find the number of books you need to multiply the number of shelves by the number of books on each shelf.

Number of books = number of shelves × number of books on each shelf
$$= 68 \times 73 = 4964$$

1   Multiply three hundred and fifty-six by twenty-three.

2   One jar of marmalade weighs 454 grams. Find the weight of 24 jars.

3   On a school outing 8 coaches were used, each taking 34 children. How many children went on the school outing?

4   A school hall has 30 rows of seats. Each row has 28 seats. How many seats are there?

5   Find the value of one hundred and fifty multiplied by itself.

6   A car park has 34 rows and each row has 42 parking spaces. How many cars can be parked?

7   A supermarket takes delivery of 54 crates of soft drink cans. Each crate contains 48 cans. How many cans are delivered?

8   A school day is 7 hours long. How many minutes are there in the school day?

9   A block of flats has 44 storeys. Each storey has 18 flats. How many flats are there in the block?

10 A lightbulb was tested by being left on non-stop. It failed after 28 days exactly. For how many hours was it working?

Division of whole numbers

$36 \div 8$ means 'How many eights are there in 36?' We can find out by repeatedly taking 8 away from 36:

$$36 - 8 = 28$$
$$28 - 8 = 20 \qquad \text{So there are 4 eights in 36 with 4 left over.}$$
$$20 - 8 = 12$$
$$12 - 8 = 4$$

Thus $\qquad 36 \div 8 = 4, \quad$ remainder 4.

A quicker way uses the multiplication facts.

We know that $32 = 4 \times 8$

therefore $36 \div 8 = 4$, remainder 4

To find $534 \div 3$ start with the hundreds:

5 (hundreds) $\div 3 = \underline{1}$ (hundred), remainder 2 (hundreds)

Take the remainder, 2 (hundreds), with the tens:

23 (tens) $\div 3 = \underline{7}$ (tens), remainder 2 (tens)

Take the remainder, 2 (tens), with the units:

24 (units) $\div 3 = \underline{8}$ units

Therefore $534 \div 3 = 178$

This can be set out as:

$$\begin{array}{r} 1\ 7\ 8 \\ 3\overline{)5^2 3^2 4} \end{array}$$

Exercise 2f

Find $4669 \div 5$

$$\begin{array}{r} 9\ 3\ 3 \quad \text{r. } 4 \\ 5\overline{)46^1 6^1 9} \end{array}$$

$4669 \div 5 = 933$, r. 4

First divide 5 into 46. Put the answer (9) above the 6 and carry 1 that is left over as shown; next 5 into 16 goes 3 times with 1 left over, so put 3 next to the 9 in the answer. Carry the 1. Finally, 5 into 19 goes 3 times with 4 left over. Put 3 above the 9. The 4 left over is the remainder.

Do the following calculations and give the remainder where there is one:

1	$87 \div 3$	**7**	$73 \div 5$	**13**	$39 \div 3$	**19**	$605 \div 3$	**25**	$192 \div 8$
2	$56 \div 4$	**8**	$83 \div 4$	**14**	$21 \div 9$	**20**	$497 \div 4$	**26**	$294 \div 9$
3	$36 \div 6$	**9**	$69 \div 3$	**15**	$78 \div 6$	**21**	$855 \div 5$	**27**	$570 \div 7$
4	$57 \div 3$	**10**	$82 \div 6$	**16**	$54 \div 2$	**22**	$693 \div 3$	**28**	$680 \div 8$
5	$72 \div 4$	**11**	$78 \div 8$	**17**	$639 \div 3$	**23**	$721 \div 7$	**29**	$731 \div 6$
6	$97 \div 2$	**12**	$85 \div 7$	**18**	$548 \div 2$	**24**	$358 \div 5$	**30**	$702 \div 5$

31	$3501 \div 3$	**34**	$1758 \div 5$	**37**	$7399 \div 5$	**40**	$2009 \div 7$	**43**	$2481 \div 7$
32	$1763 \div 4$	**35**	$3852 \div 9$	**38**	$8772 \div 4$	**41**	$1788 \div 9$	**44**	$6910 \div 4$
33	$4829 \div 2$	**36**	$6405 \div 6$	**39**	$9712 \div 8$	**42**	$1098 \div 6$	**45**	$7505 \div 5$

(?) Puzzle

Write each of the digits from 1 to 9, one in each box, so that all three expressions below are correct. Each digit is therefore used once and once only.

□ ÷ □ = □

□ + □ = □

□ − □ = □

Division by 10, 100, 1000, …

$812 \div 10$ means 'How many tens are there in 812?'

There are 81 tens in 810 so

$$812 \div 10 = 81, \quad \text{remainder } 2$$

$2578 \div 100$ means 'How many hundreds are there in 2578?'

There are 25 hundreds in 2500 so

$$2578 \div 100 = 25, \quad \text{remainder } 78$$

Exercise 2g

Calculate the following and give the remainder:

1	$256 \div 10$	**5**	$4910 \div 1000$	**9**	$9426 \div 1000$
2	$87 \div 10$	**6**	$57 \div 10$	**10**	$8512 \div 100$
3	$196 \div 100$	**7**	$186 \div 10$	**11**	$3077 \div 100$
4	$2783 \div 100$	**8**	$2781 \div 10$	**12**	$5704 \div 1000$

Long division

To find $2678 \div 21$ we can set the working out as follows:

```
      127
21)2678
   21↓|
   ‾‾|
   57|
   42↓
   ‾‾
  158
  147
  ‾‾‾
   11
```

There is <u>1</u> twenty-one in 26,   r. 5 (hundreds).

There are <u>2</u> twenty-ones in 57,   r. 15 (tens).

There are <u>7</u> twenty-ones in 158,   r. 11 (units).

So $2678 \div 21 = 127$,   r. 11.

Exercise 2h

Do the following calculations and give any remainders.

If you use your calculator to check your answers, it will give the whole number part of the answer but it will not give the remainder as a whole number.

1	$254 \div 20$	16	$6841 \div 15$	31	$8013 \div 40$	46	$350 \div 17$
2	$685 \div 13$	17	$2943 \div 23$	32	$2094 \div 32$	47	$724 \div 36$
3	$739 \div 41$	18	$2694 \div 31$	33	$5009 \div 60$	48	$2390 \div 56$
4	$862 \div 25$	19	$1875 \div 25$	34	$6312 \div 43$	49	$829 \div 106$
5	$394 \div 19$	20	$3621 \div 30$	35	$4321 \div 56$	50	$5241 \div 201$
6	$267 \div 32$	21	$7514 \div 34$	36	$7974 \div 17$	51	$3689 \div 151$
7	$875 \div 25$	22	$5829 \div 43$	37	$103 \div 35$	52	$8200 \div 250$
8	$269 \div 16$	23	$6372 \div 27$	38	$2050 \div 19$	53	$3606 \div 300$
9	$389 \div 23$	24	$8261 \div 38$	39	$5008 \div 45$	54	$8491 \div 150$
10	$298 \div 14$	25	$7315 \div 24$	40	$6100 \div 32$	55	$7625 \div 302$
11	$433 \div 15$	26	$8602 \div 15$	41	$700 \div 28$	56	$8110 \div 400$
12	$614 \div 27$	27	$3004 \div 31$	42	$4001 \div 36$	57	$3742 \div 600$
13	$2804 \div 13$	28	$1608 \div 25$	43	$3900 \div 43$	58	$8924 \div 120$
14	$7315 \div 21$	29	$7092 \div 35$	44	$2800 \div 14$	59	$6643 \div 242$
15	$8392 \div 34$	30	$2694 \div 30$	45	$600 \div 54$	60	$9260 \div 414$

Puzzle

Copy the following sets of numbers. Replace each □ with +, −, ×, or ÷
so that the calculations are correct.

1 9 □ 4 = 5   **2** 7 □ 3 = 21   **3** 28 □ 4 = 7   **4** 8 □ 2 = 4

Mixed operations of +, −, ×, ÷

When a calculation involves a mixture of the operations +, −, ×, ÷
we always do

multiplication and division first.

For example

$$2 \times 4 + 3 \times 6 = 8 + 18 \qquad \text{(multiplication first)}$$
$$= 26$$

Exercise 2i

Find $5 - 10 \times 2 \div 5 + 3$

$$5 - 10 \times 2 \div 5 + 3$$
$$= 5 - \quad 20 \quad \div 5 + 3 \qquad (\times \text{ done})$$
$$= 5 - \qquad 4 \ + 3 \qquad (\div \text{ done})$$
$$= 4$$

Find:

1 $2 + 4 \times 6 - 8$

2 $24 \div 8 - 3$

3 $6 + 3 \times 2$

4 $7 \times 2 + 6 - 1$

5 $18 \div 3 - 3 \times 2$

6 $7 + 4 - 3 \times 2$

7 $8 \div 2 + 6 \times 3$

8 $14 \times 2 \div 7 - 3 + 6$

9 $6 - 2 \times 3 + 7$

10 $5 + 4 \times 3 + 8 \div 2$

11 $7 + 3 \times 2 - 8 \div 2$

12 $5 - 4 \div 2 + 7 \times 2$

13 $6 \times 3 - 8 \times 2$

14 $9 \div 3 + 12 \div 6$

15 $12 \div 3 - 15 \div 5$

16 $9 + 3 - 6 \div 2 + 1$

17 $6 - 3 \times 2 + 9 \div 3$

18 $7 + 2 \times 4 - 8 \div 4$

19 $7 \times 2 + 8 \times 3 - 2 \times 6$

20 $5 \times 3 \times 2 - 2 \times 3 \times 4$

21 $10 \times 3 \div 15 + 6$

22 $8 + 7 \times 4 \div 2$

23	$3 \times 8 \div 4 + 7$	**32**	$7 \times 2 - 3 + 6 \div 2$
24	$9 \div 3 + 7 \times 2$	**33**	$8 + 3 \times 2 - 4 \div 2$
25	$4 - 8 \div 2 + 6$	**34**	$7 \times 2 - 4 \div 2 + 1$
26	$5 \times 4 \div 10 + 6$	**35**	$6 + 8 \div 4 + 2 \times 3 \times 4$
27	$6 \times 3 \div 9 + 2 \times 4$	**36**	$5 \times 3 \times 4 \div 12 + 6 - 2$
28	$7 + 3 \times 2 \div 6$	**37**	$5 + 6 \times 2 - 8 \div 2 + 9 \div 3$
29	$8 \div 4 + 6 \div 2$	**38**	$7 - 9 \div 3 + 6 \times 2 - 4 \div 2$
30	$12 \div 4 + 3 \times 2$	**39**	$9 \div 3 - 2 + 1 + 6 \times 2$
31	$19 + 3 \times 2 - 8 \div 2$	**40**	$4 \times 2 - 6 \div 3 + 3 \times 2 \times 4$

? Puzzle

1 Copy the following sets of numbers. Replace each □ with +, −, ×, or ÷ so that the calculations are correct.

   a $5 \square 4 \square 6 = 3$ **b** $8 \square 3 \square 4 = 1$

   c $3 \square 4 \square 2 = 9$ **d** $2 \square 1 \square 3 = 6$

2 Solve this cross-number puzzle.

Across	**1**	$127 - 64$
	3	$44 + 73 - 58$
	4	6×53
	5	330×41
	11	$9 \times 10 - 9$
	12	The next number after 40
Down	**2**	$3 \times 13 - 6$
	3	$464 \div 8$
	5	$625 \div 5$
	6	74×7
	7	9×89
	9	$5 \times 8 - 27 \div 3$
	10	$2 \times 19 - 4$

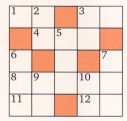

Using brackets and order of operations

If we need to do some addition and/or subtraction before multiplication and division we use brackets round the section that is to be done first.

For example $2 \times (3 + 2)$ means work out $3 + 2$ first.

So
$$2 \times (3 + 2) = 2 \times 5$$
$$= 10$$

For a calculation with brackets and a mixture of ×, ÷, + and − we first work out the inside of the Brackets, then we do the Multiplication and Division, and lastly the Addition and Subtraction.

The capital letters in the last sentence are the same as those in the following sentence:

Bless My Dear Aunt Sally

This should help you remember the order of working.

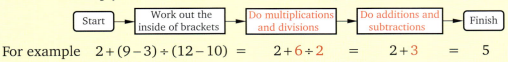

For example $\quad 2+(9-3)\div(12-10) \;=\; 2+6\div 2 \;=\; 2+3 \;=\; 5$

Exercise 2j

Find $2\times(3\times 6-4)+7-12\div 6$

$$
\begin{aligned}
2\times(3\times 6-4)+7-12\div 6 &= 2\times(18-4)+7-12\div 6 \\
&= 2\times 14+7-12\div 6 \quad \text{(inside bracket first)} \\
&= 28+7-2 \quad\quad\;\; \text{(× and ÷ next)} \\
&= 33 \quad\quad\quad\quad\;\; \text{(lastly + and −)}
\end{aligned}
$$

Find:

1. $12\div(5+1)$
2. $8\times(3+4)$
3. $(5-2)\times 3$
4. $(6+1)\times 2$
5. $(3+2)\times(4-1)$
6. $(3-2)\times(5+3)$
7. $7\times(12-5)$
8. $(6+2)\div 4$
9. $(8+1)\times(2+3)$
10. $(9-1)\div(6-2)$
11. $2+3\times(3+2)$
12. $7-2\times(5-3)$
13. $8-5+2\times(4+3)$
14. $2\times(7-2)\div(16-11)$
15. $4+3\times(2-1)+8\div(9-7)$

16. $6\div(10-8)+4$
17. $7\times(12-6)-12$
18. $12-8-3\times(9-8)$
19. $4\times(15-7)\div(17-9)$
20. $5\times(8-2)+3\times(7-5)$
21. $6\times 8-18\div(2+4)$
22. $10\div 5+20\div(4+1)$
23. $5+(2\times 10-5)-6$
24. $8-(15\div 3+4)+1$
25. $(2\times 3-4)+(33\div 11+5)$
26. $(18\div 3+3)\div(4\times 4-7)$
27. $(50\div 5+6)-(8\times 2-4)$
28. $(10\times 3-20)+3\times(9\div 3+2)$
29. $(7-3\times 2)\div(8\div 4-1)$
30. $(5+3)\times 2+(40\div 8-3)$

Exercise 2k

In the market I bought 3 oranges that cost $25 each and one cabbage that cost $85. I paid with a $500 note. How much change did I get?

First you need to find the cost of the oranges.

Cost of oranges = number of oranges × cost of one orange

$$= \$(3 \times 25) = \$75$$

Cost of the cabbage = $85

Total cost = cost of oranges + cost of cabbage

$$= \$75 + \$85 = \$160$$

To find the change you need to find the difference between $500 and $160.

Therefore the change from $500 is $500 − $160

$$= \$340$$

1 How many apples costing $80 each can I buy with $500?

2 I bought 5 oranges that cost $30 each and 2 lemons that cost $15 each. How much did I spend?

3 If a bus holds 30 children, how many buses are needed to take 420 children on a school outing?

4 Three children went into a sweet shop. The first child bought three sweets costing $5 each, the second child bought three sweets costing $4 each and the third child bought three sweets costing $6 each. How much money did they spend together?

5 A girl saves the same amount each week. After 8 weeks she has $96. How much does she save each week?

6 I bought five stamps at $17 each. How much change did I get from $100?

7 A car travelling at 50 miles an hour took 3 hours to travel from Kingston to Mandeville. How many miles did the car travel?

8 A club started the year with 82 members. During the year 36 people left and 28 people joined. How many people belonged to the club at the end of the year?

9 One money box has five $5 coins and four $10 coins in it. Another money box has six $10 coins and ten $20 coins in it. What is the total sum of money in the two money boxes?

10   A grocer bought a sack of potatoes weighing 50 kg. He divided the potatoes into bags, so that each bag held 3 kg of potatoes. How many complete bags of potatoes did he get from his sack?

11   At a school election one candidate got 26 votes, and the other candidate got 35 votes. 10 voting papers were spoiled and 5 pupils did not vote. How many children could have voted altogether?

12   I bought three pencils costing $18 each. How much change did I get from $100?

13   Three children are given $600 to split equally amongst them. How much does each child get?

14   A man can walk up a flight of steps at the rate of 30 steps a minute. It takes him 3 minutes to reach the top. How many steps are there?

15   An extension ladder is made of three separate parts, each 300 cm long. There is an overlap of 30 cm at each junction when it is fully extended. How long is the extended ladder?

16   Jane, Sarah and Claire come to school with $20 each. Jane owes Sarah $10 and she also owes Claire $5. Sarah owes Jane $4 and she also owes Claire $8. When all their debts are settled, how much money does each girl have?

17   At the bookstore I buy two comics costing $84 each, and a magazine costing $95. How much change do I get from $500?

18   A man gets paid $20 000 for a five day working week. How much does he get paid a day?

19   The total number of children in the first year of a school is 500. There are 50 more girls than boys. How many of each are there?

20   4000 apples are packed into boxes, each box holding 75 apples. How many boxes are required?

21   In a book of street plans of a town, the street plans start on page 6 and end on page 72. How many pages of street plans are there?

22   My great-grandmother died in 1894, aged 62. In which year was she born?

23   How many times can 5 be taken away from 132?

24   A palm tree was planted in the year in which Sir Grantley Adams was born. He died in 1971, aged 73. How old was the palm tree in 1984?

25   A mountaineer starts from a point that is 150 m above sea level. He climbs 200 m and then descends 50 m before climbing another 300 m. How far is he now above sea level?

26 A bus leaves the bus station at 9.30 a.m. It reaches the Town Hall at 9.40 a.m. and gets to the railway station at 9.52 a.m. How long does it take to go from the Town Hall to the railway station?

27 A class is told to work out the odd-numbered questions in an exercise containing 30 questions. How many questions do they have to do?

28 In the hardware shop I bought 3 screws that cost $15 each and 2 light bulbs that cost $95 each. I paid with a $500 note. How much change did I get?

29 A vegetable plot is 1000 cm long. Cabbages are planted in a row down the length of the plot. If the cabbages are planted 30 cm apart and the first cabbage is planted 5 cm from the end, how many cabbages can be planted in one row?

30 An airport timetable reads as follows:

Dominica	depart	9.30 a.m.
Guadeloupe	arrive	10.30 a.m.
	depart	11.35 a.m.
Antigua	arrive	12.00 p.m.

How long does the journey from Dominica to Guadeloupe take?

How long does the journey from Guadeloupe to Antigua take?

Number patterns

Exercise 21

2	7	6
9	5	1
4	3	8

This is a magic square.
The numbers in every row, in every column and in each diagonal add up to 15.

Copy and complete the following magic squares. Use the numbers 1 to 9 just once in each, and use a pencil in case you need to rub out!

1

8		
	5	
4		2

2

4	9	
	5	
	1	

3 Use each of the numbers 1 to 16 just once to complete the 4×4 magic square on the right. Each row, column and diagonal should add up to 34.

		7	
15			
9	5	16	
8		1	13

4   Make up a 3×3 magic square of your own. Use the numbers
1 to 9 just once each and put 5 in the middle.

This is a magic triangle.

The sum of the numbers along each side is equal to 9.

In the next questions, use the numbers 1, 2, 3, 4, 5 and 6 to fill in the
circles so that the sum along each side is equal to the number given
below the diagram. Each number is to be used exactly once in each
question.

5

10

6

11

7

12

8   Using the results in **5**, **6** and **7**:

(i)   Complete the table.
(ii) What is the sum, t, of the
      digits used in each triangle?
(iii) What do you notice about
      t and $3s - v$?

Sum along each side, s	10	11	12
Sum of vertices, v			
$3s - v$			

9   1, 3, 5, 7, ...   This is a sequence. By looking at it you should be
                         able to find the rule for getting the next number.

In this sequence, the next number is always 2 bigger than the number before it.

Write down the next two numbers in this sequence.

10 1, 4, 7, 10,   **14** 3, 6, 9, 12,   **18** 1, 10, 100, 1000,

11 12, 10, 8, 6,   **15** 64, 32, 16, 8,   **19** 81, 72, 64, 54,

12 1, 5, 9, 13,   **16** 1, 3, 9, 27,   **20** 3, 7, 11, 15,

13 2, 4, 8, 16,   **17** 4, 9, 16, 25,   **21** 5, 10, 17, 26,

Investigation

1 Consider the following pattern:

$$1 \qquad\qquad = 1 = 1 \times 1$$
$$1+3 \qquad = 4 = 2 \times 2$$
$$1+3+5 \qquad = 9 = 3 \times 3$$
$$1+3+5+7 = 16 = 4 \times 4$$

Write down the next three lines in this pattern.

Now try and write down (without adding them up) the sum of
a the first eight odd numbers   **b** the first twenty odd numbers.

2 Consider the following pattern:

$$2 \qquad\qquad = 2 = 1 \times 2$$
$$2+4 \qquad = 6 = 2 \times 3$$
$$2+4+6 \qquad = 12 = 3 \times 4$$
$$2+4+6+8 = 20 = 4 \times 5$$

Write down the next three lines in this pattern.

How many consecutive even numbers, beginning with 2, have a sum of
156? ($156 = 12 \times 13$).

3 Try to find the pattern in the given triangle of numbers. Can
you write down the next three rows? Do you know that this
triangle has a special name? Perhaps your teacher may
help you to find this name.

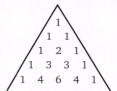

4 Fifteen red snooker balls are placed in the frame
as shown. A second layer is then placed on top
so that they rest on these in spaces marked
with crosses. This is followed by more layers
until there is a single ball at the top of
the pyramid. How many balls are needed to make
this pyramid?

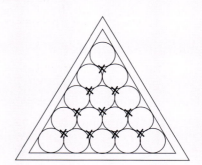

5 Consider this pattern:

$$1 \qquad\qquad\qquad = 1$$
$$1-3+5 \qquad\qquad = 3$$
$$1-3+5-7+9 \qquad = 5$$
$$1-3+5-7+9-11+13 = 7$$

Write down the next three lines in this pattern.

6 This is a class game.
Start with a number and then each pupil in turn adds on a fixed number to the last number called. For example, if you start with 5 and each pupil adds on 4 to the last number called, it will go 5, 9, 13, 17, ... If you make a mistake you are out. Ask your teacher for the name of this pattern of numbers.

7 A different version of the game in question **6** is to start with a fairly high number and then each pupil in turn subtracts a fixed number from the last one called.

Types of number

Square numbers

A square number can be represented by a number of dots arranged in a square.

For example, 25 is a square number because 25 dots can be arranged as a 5×5 square.

$25 = 5 \times 5$

The smallest square number is 1 because $1 = 1 \times 1$.

Rectangular numbers

Any number that can be shown as a rectangular pattern of dots is called a rectangular number. For example, 24 is a rectangular number because 24 dots can be arranged as

or

5 is not a rectangular number because a line of dots is not a rectangle.

Triangular numbers

A triangular number can be shown as dots arranged in rows so that each row is one dot longer than the row above.

These are the first four triangular numbers:

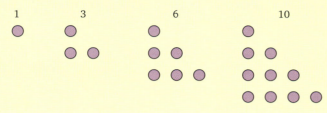

Exercise 2m

1   Which of the following numbers are square numbers?
    4, 6, 8, 9, 12, 18, 30, 36, 40

2   Which of the following numbers are rectangular numbers?
    6, 8, 11, 14, 15

    Give a reason for your answer.

3   Show that 12 is a rectangular number in two different ways.

4   Show that 18 is a rectangular number in two different ways.

5   Show that 36 is a rectangular number in three different ways.

6   Draw dot patterns for the next three triangular numbers after 10.
    Write down the next three triangular numbers after 10.

7   Without drawing dot patterns, write down the next three triangular
    numbers after 28.

8   Look at the pattern.

$$
\begin{array}{ccccccc}
& & & 1 & & & \\
& & 1 & 2 & 1 & & \\
& 1 & 2 & 3 & 2 & 1 & \\
1 & 2 & 3 & 4 & 3 & 2 & 1 \\
\end{array}
$$

    What total do you get for each line in this pattern?
    Are all these totals rectangular numbers and/or square numbers?

9   Repeat question **8** for the pattern formed by adding the odd numbers.

$$
\begin{array}{l}
1 \\
1+3 \\
1+3+5 \\
1+3+5+7
\end{array}
$$

10 What type of number do you get by adding the numbers in each row of this pattern?

$$1$$
$$1 + 2$$
$$1 + 2 + 3$$
$$1 + 2 + 3 + 4$$
$$1 + 2 + 3 + 4 + 5$$

11 Write down the numbers between 1 and 12 that are
   a square numbers   **b** rectangular numbers   **c** triangular numbers.

12 Which of the numbers between 24 and 40 are
   a square numbers   **b** rectangular numbers   **c** triangular numbers?

Mixed exercises

Exercise 2n

Find:

1 $126 + 501 + 378$ **3** 76×9 **5** $350 + 8796 - 2538$ **7** $35 + 86 + 94 + 27$

2 $153 - 136$ **4** $84 \div 3$ **6** $8 \times 321 - 1550$ **8** $20 \div (9 - 4) + 3$

9 How many packets of popcorn costing \$45 each can I buy with \$100?

10 I buy three bars of chocolate costing \$28 each. How much change do I get from \$100?

Exercise 2p

Find:

1 $92 + 625 + 153$ **3** 84×8 **5** $(7 + 30) \times 2 - 45$ **7** $68 - 42 + 12 \times 2$

2 $247 - 193$ **4** $79 \div 8$ **6** $382 - 792 \div 3$ **8** $79 - 35 + 56 - 63$

9 How many times can 6 be taken away from 45?

10 The contents of a tin of sweets weigh 2500 grams. The sweets are divided into packets each weighing 500 grams. How many packets of sweets can be made up?

Exercise 2q

Find:

1. $296 + 1025 + 983$
2. $347 - 84$
3. 7×59
4. 106×32
5. $2501 \div 9$
6. $7863 \div 20$
7. $940 + 360 - 1040$
8. $2983 \div 150$

9. A youth club has 80 members. There are 10 more boys than girls. How many of each are there?

10. There were two candidates in a school election and they got 25 votes and 32 votes. 10 voting papers were spoiled. If 100 children could have voted, how many children did not vote?

Exercise 2r

Find:

1. $749 + 821 + 1563$
2. $278 - 109$
3. 205×40
4. 284×16
5. $2781 \div 10$
6. $728 - 180 \div 12$
7. $15 - 4 \times (12 - 9)$
8. $54 + (7 \times 8 - 10) + 32$

9. I buy three stamps costing $21 each. How much change do I get from $100?

10. Add the number of small triangles in each diagram.

     What type of number are they?

In this chapter you have seen that...

✔ when you are working in columns for long multiplication and in division you should keep units under units, tens under tens, and so on

✔ brackets are used to show what needs to be done first then do multiplication and division before addition and subtraction

✔ square numbers can be shown as a square pattern of dots

✔ rectangular numbers can be shown as a rectangular pattern of dots.

3 Sets

Did you know?

Charles Dodgson (1832–1899), who is better known as
Lewis Carroll, the author of *Alice in Wonderland*, was an
Oxford mathematician who did a lot of work on sets.

Key words

disjoint sets, element, empty set, equal set, finite set, infinite set,
intersection of sets, member, null set, set, subset, union of sets, universal
set, Venn diagram, the symbols $\in$, $\notin$, $\cup$, $\varnothing$, { }, $\cup$, $\cap$.

The branch of mathematics known as Set Theory was founded by Georg Cantor.

Set notation

A *set* is a clearly defined collection of things that have something in common. We talk about a set of drawing instruments, a set of cutlery and a set of books.

Name some sets.

Things that belong to a set are called *members* or *elements*. These members or elements are usually separated by commas and written down between curly brackets or braces { }.

Instead of writing   'the set of musical instruments'

we write   {musical instruments}

Exercise 3a

1   Use the correct set notation to write down
   a   The set of foreign cars.
   b   The set of pupils in my class.
   c   The set of subjects I study at school.
   d   The set of furniture in this room.

> You can write down any two members you can think of. For example {Imported cars} includes any model you can think of such as Honda Civic, and so on.

2   Write down two members from each of the sets given in question **1**.

Describing members

We do not have to list all the members of a set; frequently we can use words to describe the members in a set.

For example, instead of {Sunday, Monday, ..., Saturday} we could say {the days of the week} and instead of {5, 6, 7, 8, 9} we could say {whole numbers from 5 to 9 inclusive}.

We could write {a, b, c, d, e} = {the first five letters of the alphabet}.

Exercise **3b**

In questions **1** to **10** describe in words the given sets:

  1   {w, x, y, z}

  2   {January, June, July}

  3   {June, July, August}

  4   {Grenada, St Vincent, St Lucia, Dominica}

  5   {Plymouth, St John's, Basseterre}

  6   {2, 4, 6, 8, 10, 12}

  7   {1, 2, 3, 4, 5, 6}

  8   {2, 3, 5, 7, 11, 13}

  9   {45, 46, 47, 48, 49, 50}

  10   {15, 20, 25, 30, 35}

In questions **11** to **15** describe a set which includes the given members and state another member of it:

  11   {Peter, John, David, Richard}

  12   {overcoat, raincoat, sweater, windcheater}

  13   {rice puffs, corn flakes, bran flakes, muesli}

  14   {hibiscus, croton, bougainvillea}

  15   {*Macbeth, Julius Caesar, King Lear, Romeo and Juliet*}

In the remaining questions list the members in the given sets:

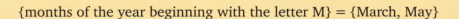

{months of the year beginning with the letter M} = {March, May}

  16   {whole numbers greater than 10 but less than 16}

  17   {the first eight letters of the alphabet}

  18   {the letters used in the word 'mathematics'}

  19   {the four main islands forming the Windward group}

  20   {the four main islands forming the Leeward group}

  21   {subjects I study}

  22   {oceans of the world}

  23   {foods I ate for breakfast this morning}

  24   {prime numbers less than 20}

  25   {even numbers less than 20}

A prime number cannot be divided by any number other than itself and one.

26 {odd numbers between 20 and 30}

 27 {multiples of 3 between 10 and 31}

28 {multiples of 7 between 15 and 50}

29 {capital cities in the Windward Islands}

30 {Caricom states}

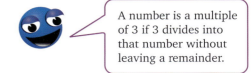

A number is a multiple of 3 if 3 divides into that number without leaving a remainder.

? Puzzle

Write down a set using five odd digits whose sum is 14.

The symbol ∈

Instead of writing

'August is a member of the set of months of the year'

we write   August ∈ {months of the year}

The symbol ∈ means 'is a member of' or 'is an element of'.

Exercise 3c

Write the following statements in set notation.

1   Apple is a member of the set of fruit.

2   Shirt is a member of the set of clothing.

3   Dog is a member of the set of domestic animals.

4   Geography is a member of the set of school subjects.

5   Carpet is a member of the set of floor coverings.

6   Hairdressing is a member of the set of occupations.

The symbol ∉

We are all aware that August is *not* a member of the set of days of the week.

Since we have chosen ∈ to mean 'is a member of' we use ∉ to mean 'is *not* a member of'. We can therefore write

'August is not a member of the set of days of the week'

as   August ∉ {days of the week}

Exercise 3d

Write the following statements in set notation:

1 Orange is not a member of the set of animals.

2 Cat is not a member of the set of fruit.

3 Table is not a member of the set of trees.

4 Shirt is not a member of the set of subjects I study.

5 Anne is not a member of the set of boys' names.

6 Chisel is not a member of the set of buildings.

7 Cup is not a member of the set of bedroom furniture.

8 Mercedes is not a member of the set of Japanese cars.

9 Aeroplane is not a member of the set of foreign countries.

10 Curry is not a member of the set of breeds of dogs.

Now write each of the following in set notation:

11 Porridge is a member of the set of breakfast cereals.

12 Electricity is not a member of the set of building materials.

13 Water is not a member of the set of metals.

14 Spider is a member of the set of living things.

15 Saturday is a member of the set of days of the week.

16 A snapper is a fish.

17 August is not the name of a day of the week.

18 Spain is a European country.

19 Brazil is not an Asian country.

Write down the meaning of:

20 Football $\in$ {team games}

21 Shoes $\notin$ {beverages}

22 Hockey $\notin$ {electrical appliances}

23 Needle $\in$ {metal objects}

24 Susan $\notin$ {boys' names}

25 Using the correct notation write down
    a three members that belong to
    b three members that do not belong to

$\qquad$ {dairy produce}

26 Using the correct notation, write down
   a three members that belong to
   b three members that do not belong to
                    {clothes}

Finite and infinite sets

Frequently, we need to refer to a set several times. When this is so we label the set with a capital letter. For example

$$A = \{\text{months of the year beginning with the letter J}\}$$

or $$A = \{\text{January, June, July}\}$$

In many cases it is not possible to list all the members of a set. When this is so we write down the first few members followed by dots.

For example, if $N = \{\text{positive whole numbers}\}$

we could write $N = \{1, 2, 3, 4, ...\}$

Similarly if $X = \{\text{even numbers}\}$ and $Y = \{\text{odd numbers}\}$

we could write $X = \{2, 4, 6, 8, ...\}$ and $Y = \{1, 3, 5, 7, ...\}$

Sets like N, X and Y are called *infinite sets* because there is no limit to the number of members each contains. When we can write down, or count, all the members in a set, the set is called a *finite set*.

Equal sets

When two sets have exactly the same members they are said to be equal.

If $A = \{2, 4, 6, 8\}$ and $B = \{6, 4, 8, 2\}$

then $A = B$

Similarly, if $X = \{\text{prime numbers less than 8}\}$
$= \{2, 3, 5, 7\}$

and $Y = \{\text{whole numbers up to 7 inclusive, except 1, 4 and 6}\}$
$= \{2, 3, 5, 7\}$

then $X = Y$

The order in which the members are listed does not matter, neither does the way in which the sets are described.

Exercise 3e

Determine whether or not the following sets are equal:

1   A = {chair, table, desk, blackboard}

   B = {desk, blackboard, table, chair}

2   X = {d, i, k, f, w}

   Y = {f, w, k, i}

3   V = {4, 6, 8, 10, 12}

   W = {even numbers from 4 to 12 inclusive}

4   C = {i, e, a, u, o}

   D = {vowels}

5   P = {Capital cities of all the Caribbean islands}

   Q = {Roseau, Kingston, Plymouth, San Fernando}

Empty set

Have you ever seen a woman with three eyes or a man with four legs? We hope not, for neither exists. There are no members in either of these sets. Such a set is called an *empty* or *null* set and is written { } or Ø.

Exercise 3f

1   Give some examples of empty sets.

2   Which of the following sets are empty?

   a   {dogs with wings}

   b   {men who have landed on the moon}

   c   {children more than 5 m tall}

   d   {cars that can carry 100 people}

   e   {men more than 100 years old}

   f   {dogs without tails}

Subsets

If A = {Paul, Peter, John, Mary, Jane} and B = {Mary, Jane} we see that all the members of B are also members of A.

We say that B is a subset of A and write this $B \subset A$.

If X = {a, b, c} then {a, b, c}, {a, b}, {b, c}, {a, c}, {a}, {b}, {c}, and $\emptyset$ are all subsets of X.

Note that both X and $\emptyset$ are considered to be subsets of X. Subsets that do not contain all the members of X are called *proper subsets*. All the subsets given above except {a, b, c} are therefore proper subsets of X.

Exercise 3g

1   If A = {w, x, y, z} write down all the subsets of A that have two members.

2   If B = {Anne, Bernard, Clive, Doris} write down all the subsets of B that have two female members.

3   If N = {1, 2, 3, ..., 10} list the following subsets of N:

   A = {odd numbers} B = {even numbers} C = {prime numbers}

4   Give a subset with at least three members for each of the following sets:

   a   {W.I. cricket captains after 1960}   **b**   {rivers}
   c   {oceans}   **d**   {Shell Shield cricket teams}

5   If X = {1, 2, 3, 5, 7, 11, 13} which of the following sets are proper subsets of X?

   a   {positive odd numbers less than 6}
   b   {positive even numbers less than 4}
   c   {positive prime numbers less than 14}
   d   {positive odd numbers between 10 and 14}

Universal set

Consider the set X = {natural numbers less than 16}, i.e. the set {1, 2, 3, 4, ..., 15}.

Now consider the sets A, B and C whose members are in X such that

$$A = \{\text{prime numbers}\} = \{2, 3, 5, 7, 11, 13\}$$

$B = \{\text{multiples of } 3\} = \{3, 6, 9, 12, 15\}$

$C = \{\text{multiples of } 5\} = \{5, 10, 15\}$

The original set $\{1, 2, 3, 4, ..., 15\}$ is called the *universal set* for the sets
A, B and C. It is a set that contains all the members that occur in the
sets A, B and C as well as some other members that are not
found in any of these three. The *universal set* is denoted
by the symbol U.

For example, a universal set for $\{\text{cup, plate, saucer}\}$
could be $\{\text{crockery}\}$.

We write $U = \{\text{crockery}\}$.

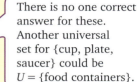

There is no one correct
answer for these.
Another universal
set for {cup, plate,
saucer} could be
$U = \{\text{food containers}\}$.

Exercise 3h

Suggest a suitable universal set for:
1   $\{8, 12, 16, 17, 20\}$
2   $\{\text{vowels}\}$
3   $\{\text{rivers in Jamaica}\}$
4   $\{\text{prefects}\}$
5   $\{\text{cats with three legs}\}$
6   $\{\text{sparrows}\}$

$U = \{\text{boys' names}\}$

Two subsets are $\{\text{John, Peter, Paul}\}$ and $\{\text{Dino, Enoch, Roshan}\}$.

Write down two subsets, each with at least two members, for each of the
following universal sets:
7   $U = \{\text{girls' names}\}$
8   $U = \{\text{European countries}\}$
9   $U = \{\text{Caribbean countries}\}$
10   $U = \{\text{Members of Parliament}\}$
11   $U = \{\text{school subjects}\}$
12   $U = \{\text{colours}\}$

Suggest a universal set for:
13   $\{\text{set squares, protractors, rulers}\}$
14   $\{\text{houses, flats, bungalows}\}$

15   {Nissan, Mercedes, Jaguar, Datsun, Subaru}

16   {trainers, shoes, sandals, boots}

17   {golfers, football players, netball players, sprinters}

Venn diagrams

Many years ago a Cambridge mathematician named John Venn (1834–1923) studied the algebra of sets and introduced the diagrams that now bear his name. In a Venn diagram the universal set (*U*) is usually represented by a rectangle and subsets of the universal set are usually shown as circles inside the rectangle. There is nothing special about circles – any convenient enclosed shape would do.

If *U* = {schoolchildren} then *A* could be {pupils in my school}, i.e. *A* is a subset of *U*.

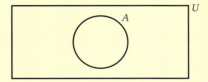

Similarly, if *B* = {pupils in the next school to my school} the diagram would be

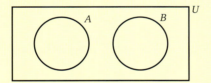

Two sets like these, which have no common members, are called *disjoint sets*

If *C* = {pupils in my class} then, because all the members of *C* are also members of *A*, i.e. *C* is a proper subset of *A*, the Venn diagram is

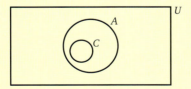

If *D* = {my school friends} the corresponding Venn diagram could be

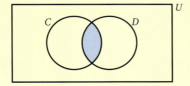

The shaded region shows the friends I have who are in my class. These friends belong to both sets. The unshaded region of *D* represents friends I have in school who are not in my class.

Union of two sets

In my class

$$A = \{\text{pupils good at maths}\} = \{\text{Frank, Javed, Asif, Sian}\}$$

and

$$B = \{\text{pupils good at French}\} = \{\text{Bina, Asif, Polly, Frank}\}$$

If the universal set is {all the pupils in my class} the names could be placed in a Venn diagram as follows:

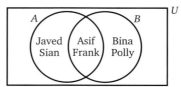

If we write down the set of all the members of my class who are good at *either* maths *or* French we have the set {Javed, Sian, Asif, Frank, Bina, Polly}. This is called the *union* of the sets *A* and *B* and is denoted by

$$A \cup B$$

Similarly if $X = \{1, 2, 3, 5\}$ and $Y = \{2, 4, 6\}$ we can illustrate these sets in the following Venn diagram:

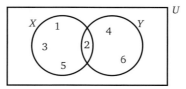

and write the union of the two sets *X* and *Y*

$$X \cup Y = \{1, 2, 3, 4, 5, 6\}$$

To find the union of two sets, write down all the members of the first set, then all the members of the second set which have not already been included.

Exercise 3i

Find the union of the two given sets in each of the following:

$A = \{3, 6, 9, 12\}$ $B = \{4, 6, 8, 10, 12\}$

6 and 12 are in both sets so they do not need to be included twice.

$A \cup B = \{3, 4, 6, 8, 9, 10, 12\}$

1   $A = \{$Peter, James, John$\}$ $B = \{$John, Andrew, Paul$\}$
2   $X = \{3, 6, 9, 12\}$ $Y = \{4, 8, 12, 16\}$
3   $P = \{$a, e, i, o, u$\}$ $Q = \{$a, b, c, d, e$\}$
4   $A = \{$a, b, c$\}$ $B = \{$x, y, z$\}$
5   $A = \{$p, q, r, s, t$\}$ $B = \{$p, r, t$\}$
6   $X = \{2, 3, 5, 7\}$ $Y = \{1, 3, 5, 7\}$
7   $X = \{5, 7, 11, 13\}$ $Y = \{6, 8, 10, 12\}$
8   $P = \{$whole numbers that divide exactly into 12$\}$
    $Q = \{$whole numbers that divide exactly into 10$\}$
9   $A = \{$letters in the word 'classroom'$\}$
    $B = \{$letters in the word 'school'$\}$
10 $P = \{$letters in the word 'arithmetic'$\}$
    $Q = \{$letters in the word 'algebra'$\}$

To represent the union of two sets in a Venn diagram we shade the combined region representing the two sets. This shaded area may occur in three ways.

a   When sets A and B have some common members.

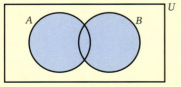

b   When A and B have no common member, i.e. when they are disjoint.

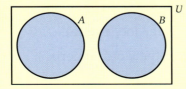

c When *B* is a proper subset of *A*.

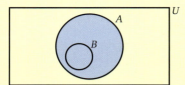

Exercise 3j

Draw suitable Venn diagrams to show the unions of the following sets:

P = {3, 6, 9, 12, 15} $\qquad$ Q = {3, 5, 7, 9, 11, 15}

3, 9 and 15 are in both sets so these go in the overlapping part.

6 and 12 are only in *P* so they go in the left-hand part of the circle marked *P*.

5, 7, and 11 go in the right-hand part of the circle marked *Q*.

The union is the combination of both sets so both circles are shaded.

$$P \cup Q = \{3, 5, 6, 7, 9, 11, 12, 15\}$$

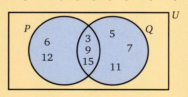

1	A = {p, q, r, s}	B = {r, s, t, u}
2	X = {1, 3, 5, 7, 9}	Y = {2, 4, 6, 8, 10}
3	P = {a, b, c, d, e, f, g}	Q = {c, d, g}
4	E = {rectangles}	F = {squares}
5	G = {even numbers}	H = {odd numbers}
6	M = {triangles}	N = {squares}
7	A = {3, 6, 9, 12, 15}	B = {4, 6, 8, 10, 12, 14}
8	P = {letters in the word 'Donald'}	Q = {letters in the word 'London'}
9	X = {Marc, Leslie, Joe, Claude}	Y = {Leslie, Sita, Joe, Yvette}
10	A = {letters in the word 'metric'}	B = {letters in the word 'imperial'}

Intersection of sets

If we return to the set of pupils in my class

$$A = \{\text{pupils good at maths}\}$$
$$= \{\text{Frank, Javed, Asif, Sian}\}$$

and

$$B = \{\text{pupils good at French}\}$$
$$= \{\text{Bina, Asif, Polly, Frank}\}$$

then Frank and Asif form the set of pupils who are good at *both* maths and French. The members that are in both sets give what is called the *intersection* of the sets A and B.

The intersection of two sets A and B is written $A \cap B$,

i.e. for the given sets, $A \cap B = \{\text{Frank, Asif}\}$

Exercise 3k

Find the intersection of $X = \{1, 2, 3, 4, 5, 6\}$ and $Y = \{1, 2, 3, 5, 7\}$

The intersection contains the elements that are in both X and Y

$$X \cap Y = \{1, 2, 3, 5\}$$

Find the intersection of the following pairs of sets:

1   $A = \{3, 6, 9, 12\}$ $B = \{5, 6, 7, 8, 9\}$
2   $X = \{4, 8, 12, 16, 20\}$ $Y = \{4, 12, 20\}$
3   $P = \{\text{Bob, Ken, Colin, Alice}\}$ $Q = \{\text{Alice, Bill, Hans, Bob}\}$
4   $C = \{o, p, q, r, s, t\}$ $D = \{a, e, i, o, u\}$
5   $A = \{\text{tomato, cabbage, apple, pear}\}$ $B = \{\text{cabbage, tomato}\}$
6   $M = \{\text{prime numbers less than 12}\}$ $N = \{\text{odd numbers less than 12}\}$
7   $P = \{4, 8, 12, 16\}$ $Q = \{8, 16, 24, 48\}$
8   $A = \{1, 2, 3, 4, 6, 12\}$ $B = \{1, 2, 5, 10\}$
9   $X = \{\text{letters in the word 'twice'}\}$ $Y = \{\text{letters in the word 'sweat'}\}$
10 $P = \{\text{letters in the word 'metric'}\}$ $Q = \{\text{letters in the word 'imperial'}\}$

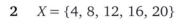

Look for the elements that are common to both sets.

Exercise 3I

Draw suitable Venn diagrams to show the intersections of the following sets:

$X = \{1, 2, 3, 4, 5, 6\}$ $Y = \{2, 3, 5, 7, 11\}$

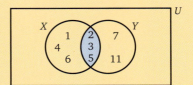

$X \cap Y = \{2, 3, 5\}$

The elements in $X \cap Y$ are in the overlap, so this is the part to shade.

1   $A = \{1, 3, 5, 7, 9, 11\}$ $B = \{2, 3, 4, 5, 6, 7\}$

2   $P = \{\text{John, David, Dino, Kay}\}$ $Q = \{\text{Pete, Dino, Omar, John}\}$

3   $X = \{a, e, i, o, u\}$ $Y = \{b, f, o, w, u\}$

4   $A = \{\text{oak, ash, elm, pine}\}$ $B = \{\text{teak, oak, sapele, elm}\}$

5   $X = \{\text{poodle, greyhound, boxer}\}$ $Y = \{\text{pug, collie, boxer, cairn}\}$

6   $P = \{4, 8, 12, 16\}$ $Q = \{8, 16, 24, 48\}$

7   $A = \{1, 2, 3, 4, 6, 12\}$ $B = \{1, 2, 4, 5, 10, 20\}$

8   $X = \{\text{letters in the word 'think'}\}$ $Y = \{\text{letters in the word 'flint'}\}$

9   $A = \{\text{letters in the word 'arithmetic'}\}$ $B = \{\text{letters in the word 'geometry'}\}$

10   $P = \{\text{prime numbers less than 10}\}$ $Q = \{\text{odd numbers less than 15}\}$

 Investigation

Consider the following sets X and Y.

$X = \{0, 2, 4, 6, ...\}$, $Y = \{1, 3, 5, 7, ...\}$

1   Describe X and Y in words.

2   Give the next three numbers in X and in Y.

3   Choose two numbers from X and find their sum. Is this sum a member of X? Is this always true?

4   Choose two numbers from Y and find their sum. Which set contains this sum? Is this always true?

5   When will the sum of any two whole numbers be in Y? Explain your answer with examples.

In this chapter you have seen that...

✔ a set is a collection of things that have something in common

✔ an infinite set has no limit on the number of members in it

✔ in a finite set, all the members can be counted or listed

✔ a proper subset of a set A contains none or some, but not all, of the members of A

✔ the union of two sets contains all the members of the first set together with the members of the second set that have not already been included

✔ the intersection of two sets contains the elements that are in both sets.

✔ when two sets have exactly the same members, they are said to be equal.

✔ a set which has no members is called an empty or null set and is written { } or Ø.

At the end of this chapter you should be able to...

1 List the set of factors of a given number.

2 Write down multiples of a given number.

3 Classify numbers as prime or composite.

4 Express a given number as a product of primes.

5 Find the Highest Common Factor (HCF) or Lowest Common Multiple (LCM) of a set of numbers.

6 Solve problems requiring the use of HCF or LCM of numbers.

Did you know?

Wherever you are a million is always a million (1 000 000).

However, a BILLION is not always a billion.

In the USA, 1 billion $= 1000 \times 1 000 000$

But in France and Great Britain, 1 BILLION used to be $1 000 000 \times 1 000 000$.

You need to know...

✔ your multiplication tables up to 10×10

✔ what a set is

✔ how to divide by whole numbers less than 10

Key words

composite number, digit, divisible, factor, highest common factor (HCF), lowest common multiple (LCM), multiple, natural number, prime number, product, set.

Factors

The number 2 is a *factor* of 12, since 2 will divide exactly into 12.

The number 12 may be expressed as the product of two factors in several different ways, namely:

$$1 \times 12 \qquad 2 \times 6 \qquad 3 \times 4$$

The numbers 1, 2, 3, 4, 6 and 12 will divide exactly into 12.

The set of factors of 12 is {1, 2, 3, 4, 6, 12}.

Exercise 4a

Express each of the following numbers as the product of two factors, giving all possibilities:

1	18	5	30	9	48	**13**	80	**17**	120
2	20	6	36	10	60	**14**	96	**18**	135
3	24	7	40	11	64	**15**	100	**19**	144
4	27	8	45	12	72	**16**	108	**20**	160

Exercise 4b

List the set of factors for each of the numbers in Exercise 4a.

Multiples

12 is a *multiple* of 2 since 12 contains the number 2 a whole number of times. The set of multiples of 2 is {2, 4, 6, 8, 10, 12, ...}.

Exercise 4c

1   Write down the set of multiples of 3 between 20 and 40.

2   Write down the set of multiples of 5 between 19 and 49.

3   Write down the set of multiples of 7 between 25 and 60.

4   Write down the set of multiples of 11 between 50 and 100.

5   Write down the set of multiples of 13 between 25 and 70.

Prime numbers

Some numbers can be expressed as the product of two different or unequal factors in only one way. For example, the only factors of 3 are 1 and 3 and the only factors of 5 are 1 and 5. Any number bigger than 1 that is of this type is called a *prime number*. Note that 1 is *not* a prime number.

A natural number, other than 1, which is not prime is *composite* e.g. 6 ($6 = 2 \times 3$).

Exercise 4d

1 Which of the following numbers are prime numbers?

$$2, 3, 4, 5, 6, 7, 8, 9, 10, 11, 12, 13$$

2 Write down the set of prime numbers between 20 and 30.

3 Write down the set of prime numbers between 30 and 50.

4 Which members of the following set are prime numbers?

$$\{5, 10, 19, 29, 39, 49, 61\}$$

5 Which members of the following set are prime numbers?

$$\{41, 57, 91, 101, 127\}$$

6 Are the following statements true or false?
   a All prime numbers are odd numbers.
   b All odd numbers are prime numbers.
   c All prime numbers between 10 and 100 are odd numbers.
   d The only even prime number is 2.
   e There are six prime numbers less than 10.

Divisibility tests

A number is divisible:

* by 2 if the last figure is even
* by 3 if the sum of the digits is divisible by 3
* by 5 if the last figure is 0 or 5
* by 9 if the sum of the digits is divisible by 9.

Exercise 4e

Is 1683 divisible by 3?

The sum of the digits is $1+6+8+3 = 18$, which is divisible by 3.
Therefore 1683 is divisible by 3.

1   Is 525 divisible by 3?

2   Is 747 divisible by 5?

3   Is 2931 divisible by 3?

4   Is 740 divisible by 5?

5   Is 543 divisible by 5?

6   Is 1424 divisible by 2?

7   Is 9471 divisible by 3?

8   Is 2731 divisible by 2?

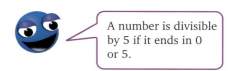

A number is divisible by 5 if it ends in 0 or 5.

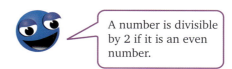

A number is divisible by 2 if it is an even number.

Is 8820 divisible by 15?

8820   is divisible by 5 since it ends in 0.

8820   is divisible by 3 since $8+8+2 = 18$ which is divisible by 3.

8820   is therefore divisible by both 5 and 3, i.e. it is divisible by 5×3 or 15.

15 is the product of the prime numbers 3 and 5, so you need to test to see if 8820 is divisible by both 3 and 5.

9   Is 10 752 divisible by 6?

10 Is 21 168 divisible by 6?

11 Is 30 870 divisible by 15?

? Puzzle

It is a curious fact that $12 \times 12 = 144$, and if the digits in all three numbers are reversed you have $21 \times 21 = 441$, which is also true.

Find other examples with this property.

Highest Common Factor (HCF)

The Highest Common Factor (HCF) of two or more numbers is the largest number that divides exactly into each of them.

For example, 8 is the HCF of 16 and 24; 15 is the HCF of 45, 60 and 120.

If you cannot see the HCF of a set of numbers, you can find it by listing all the factors of each number.

Exercise 4f

Find the HCF of 42 and 63.

Factors of 42: 1, 2, 3, 7, 14, <u>21</u>, 42
Factors of 63: 1, 3, 7, 9, <u>21</u>, 63
The HCF of 42 and 63 is 21.

> The common factors of 42 and 63 are 1, 3, 7, 21. 21 is the highest.

State the HCF of:

1	9, 12	**6**	35, 42	**11**	25, 75	**16**	25, 35, 60
2	8, 16	**7**	24, 39	**12**	22, 44	**17**	36, 52, 56
3	12, 24	**8**	18, 48	**13**	42, 84	**18**	15, 30, 45
4	14, 42	**9**	12, 30	**14**	39, 13	**19**	18, 20, 36
5	21, 28	**10**	64, 72	**15**	51, 34	**20**	6, 12, 32

 Investigation

Two or more counting numbers (1, 2, 3, …) are called *relatively prime* if their highest common factor is 1. For example, 2 and 3 are relatively prime, so are 3 and 8.

Find as many relatively prime numbers as you can less than 20.

Lowest Common Multiple (LCM)

The Lowest Common Multiple (LCM) of two or more numbers is the smallest number that divides exactly by each of the numbers.

For example, the LCM of 8 and 12 is 24 since both 8 and 12 divide exactly into 24.

Similarly the LCM of 4, 6 and 9 is 36.

If you cannot see the LCM of a set of numbers, list their multiples until you find a common multiple.

Shortcuts

The following method may also be used to find the LCM of two numbers,

e.g. find the LCM of 45 and 60.

First find the HCF of 45 and 60. This is 15.

Divide each number by the HCF, 15. We get 3 and 4.

The LCM is the product of these quotients and the HCF, i.e. $3 \times 4 \times 15 = 180$.

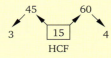

$$45 \qquad 60$$
$$3 \qquad \boxed{15} \qquad 4$$
$$\text{HCF}$$
$$\text{LCM} = 3 \times 15 \times 4 = 180$$

Exercise 4g

Find the LCM of 9 and 15.

Multiples of 9: 9, 18, 27, 36, <u>45</u>, 54, ...
Multiples of 15: 15, 30, <u>45</u>, ...
The LCM of 9 and 15 is 45.

State the LCM of:

1	3, 5	**4**	9, 12	**7**	12, 16, 24	<u>**10**</u>	18, 27, 36
2	6, 8	**5**	3, 9, 12	**8**	4, 5, 6	<u>**11**</u>	9, 12, 36
3	5, 15	**6**	10, 15, 20	<u>**9**</u>	9, 12, 18	<u>**12**</u>	6, 7, 8

Problems involving HCFs and LCMs

Exercise 4h

Mrs Walcott buys a box of chocolates for her party. She is unsure whether there will be 9 or 12 people altogether, but she is sure that whichever number it is everybody can have the same number of chocolates. What is the least number of chocolates that needs to be in the box?

You need to find the smallest number that 9 and 12 will divide into exactly, i.e. the LCM of 9 and 12.

Multiples of 9: 9, 18, 27, <u>36</u>, 45, ...
Multiples of 12: 12, 24, <u>36</u>, ...
The smallest number that 9 and 12 will divide into exactly is therefore 36.
Therefore 36 is the least number of chocolates.

1 What is the smallest sum of money that can be made up of an exact number of 10 c coins or of 25 c coins?

2 Find the least sum of money into which $24, and $54 will divide exactly.

3 Find the smallest length that can be divided exactly into equal sections of length 8 m or 12 m.

4 A room measures 450 cm by 350 cm. Find the side of the largest square tile that can be used to tile the floor without any cutting.

5 Two model trains travel around a circular two-line track. One train takes 18 seconds to go round. The other takes 15 seconds to go round. They start side by side. How many seconds will it be before they are side by side again?

6 Find the largest number of children that can equally share 72 sweets and 54 chocolates.

7 In the first year of a large high school it is possible to divide the pupils into equal sized classes of either 24 or 30 or 32 and have no pupils left over. Find the size of the smallest entry that makes this possible. How many classes will there be if each class is to have 24 pupils?

<u>8</u>   If I go up a flight of stairs two at a time I get to the top without any being left over. If I then try three at a time and again five at a time, I still get to the top without any being left over. Find the shortest flight of stairs for which this is possible. How many would remain if I were able to go up seven at a time?

? Puzzle

1   Using the four digits 2, 3, 6 and 9 once only you can make several pairs of 2-digit numbers, e.g. 26 and 93. Find 26×93.

Now pair the digits in a different way, e.g. 39 and 62 and find 39×62. What do you notice?

Can you find another four digits with the same property?

2   The church at Arima has a peal of four bells. No. 1 bell rings every 5 seconds, No. 2 bell every 6 seconds, No. 3 bell every 7 seconds and No. 4 every 8 seconds. They are first tolled together. Investigate how long it will be before they all sound together again.

! Investigation

Perfect numbers

The ancient Greeks discovered a set of numbers, each of which is equal to the sum of its factors, excluding itself.

For example, the factors of 6, excluding 6, are {1, 2, 3}.

$1 + 2 + 3 = 6$. Hence 6 is a PERFECT number.

Find some other perfect numbers. Consider all factors, not just prime factors.

Can a prime number be a perfect number? Explain your answer.

Investigate 496 to see if it is a perfect number.

> ### In this chapter you have seen that...
>
> ✔ a prime number is any number bigger than 1 whose only factors are 1 and itself
>
> ✔ you can find the largest factor that will divide exactly into a set of numbers by listing all their factors. This largest factor is called the Highest Common Factor (HCF)
>
> ✔ you can find the lowest number that all the numbers of a set will divide into exactly by writing the multiples of each number until you find one that is the same for all the numbers. This lowest number is called the Lowest Common Multiple (LCM).

5 Fractions: addition and subtraction

At the end of this chapter you should be able to...

1. Express one quantity as a fraction of another.
2. Write a fraction equivalent to a given fraction.
3. Order a set of fractions according to magnitude.
4. Add and subtract fractions.
5. Solve problems using addition and subtraction of fractions.

Did you know?

The system of writing one number above the other, as in $\frac{1}{2}$, is attributed to a Hindu mathematician, Brahmagupta. The bar between the two numbers as in $\frac{1}{2}$ was first used by the Arabs, about CE 1150.

You need to know...

✔ how to add and subtract whole numbers
✔ how to divide by a whole number
✔ what LCM means and how to find it.

Key words

cancel, common denominator, common factor, denominator, equivalent fraction, fraction, improper fraction, lowest common multiple, mixed number, numerator, proper fraction, simplify a fraction, the symbols < and >.

The meaning of fractions

Think of cutting a cake right through the middle into two equal pieces. Each piece is one half of the cake. One half is a fraction, written as $\frac{1}{2}$.

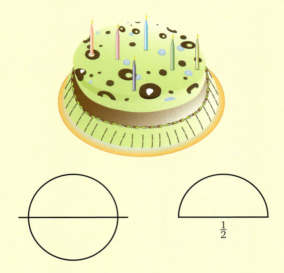

If we cut the cake into four equal pieces, each piece is one quarter, written $\frac{1}{4}$, of the cake. When one piece is taken away there are three pieces left, so the fraction that is left is three quarters, or $\frac{3}{4}$.

When the cake is divided into five equal slices, one slice is $\frac{1}{5}$, two slices is $\frac{2}{5}$, three slices is $\frac{3}{5}$ and four slices is $\frac{4}{5}$ of the cake.

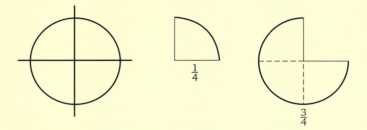

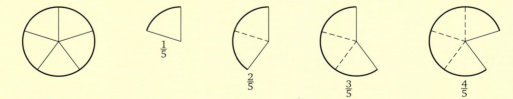

Notice that the top number in each fraction (called the *numerator*) tells you *how many* slices and the bottom number (called the *denominator*) tells you about the total number of slices that make a whole cake.

Exercise 5a

In each of the following sketches, write down the fraction that is shaded:

1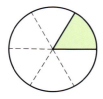

3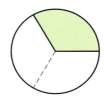

5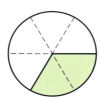

2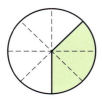

4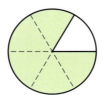

6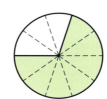

It is not only cakes that can be divided into fractions. Anything at all that can be split up can be divided into fractions.

Write down the fraction that is shaded in each of the following diagrams:

7

12

8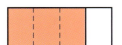

13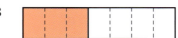

9

14

10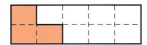

15

11

16

One quantity as a fraction of another

Quite a lot of things are divided into equal parts. For instance a week is divided into seven days, so each day is $\frac{1}{7}$ of a week. One dollar is divided into one hundred cents, so each cent is $\frac{1}{100}$ of a dollar.

Exercise 5b

In January there were 23 sunny days. What fraction of January was sunny?

There are 30 days in January, so you need to find 23 as a fraction of 31.

$$23 \text{ sunny days} = \frac{23}{31} \text{ of January}$$

1   One hour is divided into 60 minutes. What fraction of an hour is
   a   one minute
   b   nine minutes
   c   thirty minutes
   d   forty-five minutes?

2   You go to school on five days each week. What fraction of a week is this?

3   In the month of June, it rained on eleven days. What fraction of all the days in June did it rain?

In questions **4** to **13** write the first quantity as a fraction of the second quantity:

Write 10 minutes as a fraction of 1 hour.

(We must always use the same unit for both quantities. This time we will use minutes, so we want to write 10 minutes as a fraction of 60 minutes.)

$$10 \text{ minutes} = \frac{10}{60} \text{ of 1 hour}$$

4   51 days; 1 year (not a leap year)
5   $35; $100
6   $90; $500
7   35 seconds; 3 minutes
8   3 days; the month of January

9   17 days; the months of June and July together

10   5 days; 3 weeks

11   $150; $500

12   45 minutes; 2 hours

13   37 seconds; 1 hour

14   A boy gets $80 pocket money. If he spends $45, what fraction of his pocket money is left?

15   In a class of thirty-two children, ten take Spanish, eight take music and twenty-five take geography. What fraction of the children in the class take

    a   Spanish   **b**   music   **c**   geography?

16   A girl's journey to school costs $15 on one bus and $25 on another bus. What fraction of the total cost arises from each bus?

17   In an orchard there are twenty mango trees, eighteen breadfruit trees, fourteen papaya trees and ten ackee trees. What fraction of all the trees are

    a   mango trees   **b**   ackee trees   **c**   *not* papaya trees?

18   In a Youth Club with 37 members, 12 are more than 15 years old and 8 are under 14 years old. What fraction of the members are

    a   over 15   **b**   under 14   **c**   14 and over?

19   During an Easter holiday of fourteen days there were three rainy days, two cloudy days and all the other days were sunny. What fraction of the holiday was

    a   sunny   **b**   rainy?

Equivalent fractions

In the first sketch below, a cake is cut into four equal pieces. One slice is $\frac{1}{4}$ of the cake.

In the second sketch the cake is cut into eight pieces. Two slices is $\frac{2}{8}$ of the cake.

In the third sketch the cake is cut into sixteen equal slices. Four slices is $\frac{4}{16}$ of the cake.

 $\frac{1}{4}$ $\frac{2}{8}$ $\frac{4}{16}$

But the same amount of cake has been taken each time.

Therefore $\quad \frac{1}{4} = \frac{2}{8} = \frac{4}{16}$

and we say that $\frac{1}{4}$, $\frac{2}{8}$ and $\frac{4}{16}$ are *equivalent fractions*.

Now $\quad \frac{1}{4} = \frac{1\times2}{4\times2} = \frac{2}{8}$ and $\frac{1}{4} = \frac{1\times4}{4\times4} = \frac{4}{16}$

So all we have to do to find equivalent fractions is to multiply the numerator and the denominator by the same number. For instance

$$\frac{1}{4} = \frac{1\times3}{4\times3} = \frac{3}{12}$$

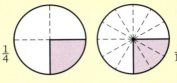

and $\quad \frac{1}{4} = \frac{1\times5}{4\times5} = \frac{5}{20}$

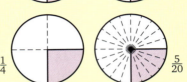

Any fraction can be treated in this way.

Exercise 5c

In questions **1** to **6** draw cake diagrams to show that:

1 $\quad \frac{1}{3} = \frac{2}{6}$ $\qquad$ **3** $\quad \frac{1}{5} = \frac{2}{10}$ $\qquad$ **5** $\quad \frac{2}{3} = \frac{6}{9}$

2 $\quad \frac{1}{2} = \frac{3}{6}$ $\qquad$ **4** $\quad \frac{3}{4} = \frac{9}{12}$ $\qquad$ **6** $\quad \frac{2}{3} = \frac{8}{12}$

Fill in the missing number to make equivalent fractions.

$\quad$ **a** $\quad \frac{1}{5} = \frac{3}{\underline{}}$

$\qquad$ If $\frac{1}{5} = \frac{3}{\underline{}}$ the numerator has been multiplied by 3 so we need to multiply the denominator by 3.

$$\frac{1}{5} = \frac{1\times3}{5\times3} = \frac{3}{15}$$

$\quad$ **b** $\quad \frac{1}{5} = \frac{\underline{}}{20}$

$\qquad$ If $\frac{1}{5} = \frac{\underline{}}{20}$ the denominator has been multiplied by 4 so you need to multiply the numerator by 4.

$$\frac{1}{5} = \frac{1\times4}{5\times4} = \frac{4}{20}$$

In questions **7** to **33** fill in the missing numbers to make equivalent fractions:

7 $\frac{1}{3} = \frac{2}{}$

8 $\frac{2}{5} = \frac{}{10}$

9 $\frac{3}{7} = \frac{9}{}$

10 $\frac{9}{10} = \frac{}{40}$

11 $\frac{1}{6} = \frac{3}{}$

12 $\frac{1}{3} = \frac{}{12}$

13 $\frac{2}{5} = \frac{6}{}$

14 $\frac{3}{7} = \frac{}{28}$

15 $\frac{9}{10} = \frac{90}{}$

16 $\frac{1}{6} = \frac{}{36}$

17 $\frac{4}{5} = \frac{}{20}$

18 $\frac{2}{3} = \frac{12}{}$

19 $\frac{2}{9} = \frac{4}{}$

20 $\frac{3}{8} = \frac{}{80}$

21 $\frac{5}{11} = \frac{}{22}$

22 $\frac{4}{5} = \frac{8}{}$

23 $\frac{1}{10} = \frac{10}{}$

24 $\frac{2}{9} = \frac{}{36}$

25 $\frac{3}{8} = \frac{}{800}$

26 $\frac{5}{11} = \frac{50}{}$

27 $\frac{4}{5} = \frac{}{50}$

28 $\frac{1}{10} = \frac{100}{}$

29 $\frac{2}{9} = \frac{20}{}$

30 $\frac{3}{8} = \frac{3000}{}$

31 $\frac{5}{11} = \frac{}{121}$

32 $\frac{4}{5} = \frac{400}{}$

33 $\frac{1}{10} = \frac{1000}{}$

Write $\frac{2}{3}$ as an equivalent fraction with denominator 24.

You need to write $\frac{2}{3}$ as $\frac{?}{24}$; 3×8 is 24 so you need to multiply the top and bottom of $\frac{2}{3}$ by 8.

$$\frac{2}{3} = \frac{2 \times 8}{3 \times 8} = \frac{16}{24}$$

34 Write each of the following fractions as an equivalent fraction with denominator 24:

a $\frac{1}{2}$ **b** $\frac{1}{3}$ **c** $\frac{1}{6}$ **d** $\frac{3}{4}$ **e** $\frac{5}{12}$ **f** $\frac{3}{8}$

35 Write each of the following fractions in equivalent form with denominator 45:

a $\frac{2}{15}$ **b** $\frac{4}{9}$ **c** $\frac{3}{5}$ **d** $\frac{1}{3}$ **e** $\frac{14}{15}$ **f** $\frac{1}{5}$

36 Find an equivalent fraction with denominator 36 for each of the following fractions:

a $\frac{3}{4}$ **b** $\frac{5}{9}$ **c** $\frac{1}{6}$ **d** $\frac{5}{18}$ **e** $\frac{7}{12}$ **f** $\frac{2}{3}$

37 Change each of the following fractions into an equivalent fraction with numerator 12:

a $\frac{1}{6}$ **b** $\frac{3}{4}$ **c** $\frac{6}{7}$ **d** $\frac{4}{5}$ **e** $\frac{2}{3}$ **f** $\frac{1}{2}$

38 Some of the following equivalent fractions are correct but two of them are wrong. Find the wrong ones and correct them by altering the numerator:

a $\frac{2}{5} = \frac{6}{15}$ **b** $\frac{2}{3} = \frac{4}{9}$ **c** $\frac{3}{7} = \frac{6}{14}$ **d** $\frac{4}{9} = \frac{12}{27}$ **e** $\frac{7}{10} = \frac{77}{100}$

Investigation

Using the numbers 1, 2, 4 and 8 write down all the fractions you can think of that are equal to or smaller than 1. Use a single number for the numerator and a single number for the denominator. You can use a number more than once in the same fraction.

1 **a** How many fractions can you find?
  b Which fractions are equivalent fractions?
  c Which fraction does not have an equivalent fraction in the list you have written down?
2 Add 16 to the list of numbers 1, 2, 4, 8 and repeat part **1**.
3 Two-digit numbers, such as 14 and 82, can be made from the digits 1, 2, 4 and 8. Use such numbers to repeat part **1**.

Ordering fractions

Which is bigger, $\frac{5}{7}$ or $\frac{2}{3}$? To compare these two fractions we change them into the *same kind* of fraction. That means we find equivalent fractions that have the same denominator. This denominator must be a number that both 7 and 3 divide into, so our new denominator is 21.

$$\frac{5}{7} = \frac{15}{21} \text{ and } \frac{2}{3} = \frac{14}{21}$$

$\frac{15}{21}$ is bigger than $\frac{14}{21}$, so $\frac{5}{7}$ is bigger than $\frac{2}{3}$.

We often use the symbol > for the words 'is bigger than'.
Using this symbol we could write

$$\frac{15}{21} > \frac{14}{21}, \text{ so } \frac{5}{7} > \frac{2}{3}.$$

Similarly we use < for 'is less than'.

Exercise 5d

In the following questions find which is the bigger fraction:

$\frac{3}{5}$ or $\frac{7}{11}$

You need to change $\frac{3}{5}$ and $\frac{7}{11}$ into equivalent fractions with the same denominator, so you need to find the LCM of 5 and 11.

$\frac{3}{5} = \frac{33}{55}$ and $\frac{7}{11} = \frac{35}{55}$ (55 divides by 5 and by 11)

$\frac{35}{55} > \frac{33}{55}$ so $\frac{7}{11}$ is the bigger fraction.

1	$\frac{1}{2}$ or $\frac{1}{3}$	7	$\frac{2}{5}$ or $\frac{3}{7}$	13	$\frac{1}{4}$ or $\frac{3}{11}$	**19**	$\frac{2}{9}$ or $\frac{3}{11}$
2	$\frac{3}{4}$ or $\frac{5}{6}$	8	$\frac{5}{6}$ or $\frac{3}{5}$	14	$\frac{5}{7}$ or $\frac{3}{5}$	**20**	$\frac{5}{7}$ or $\frac{7}{9}$
3	$\frac{2}{3}$ or $\frac{4}{5}$	**9**	$\frac{3}{8}$ or $\frac{1}{5}$	15	$\frac{3}{8}$ or $\frac{5}{11}$	**21**	$\frac{9}{11}$ or $\frac{7}{9}$
4	$\frac{2}{9}$ or $\frac{1}{7}$	**10**	$\frac{4}{5}$ or $\frac{6}{7}$	16	$\frac{3}{10}$ or $\frac{4}{11}$	**22**	$\frac{2}{5}$ or $\frac{1}{3}$
5	$\frac{2}{7}$ or $\frac{3}{8}$	**11**	$\frac{3}{5}$ or $\frac{4}{7}$	**17**	$\frac{1}{4}$ or $\frac{2}{7}$	23	$\frac{4}{7}$ or $\frac{3}{5}$
6	$\frac{2}{3}$ or $\frac{3}{4}$	12	$\frac{3}{4}$ or $\frac{2}{3}$	18	$\frac{5}{8}$ or $\frac{4}{7}$	**24**	$\frac{5}{8}$ or $\frac{6}{11}$

In questions **25** to **36**, put either > or < between the fractions:

25	$\frac{1}{4}$	$\frac{2}{7}$		**31**	$\frac{3}{5}$	$\frac{2}{3}$
26	$\frac{2}{3}$	$\frac{5}{8}$		**32**	$\frac{2}{9}$	$\frac{1}{5}$
27	$\frac{3}{7}$	$\frac{1}{2}$		**33**	$\frac{4}{9}$	$\frac{5}{11}$
28	$\frac{5}{8}$	$\frac{7}{10}$		**34**	$\frac{2}{11}$	$\frac{1}{7}$
29	$\frac{3}{10}$	$\frac{1}{4}$		**35**	$\frac{8}{11}$	$\frac{3}{4}$
30	$\frac{1}{3}$	$\frac{2}{5}$		**36**	$\frac{7}{8}$	$\frac{7}{9}$

You need to change $\frac{1}{4}$ and $\frac{2}{7}$ into equivalent fractions with the same denominator, then you can see if $\frac{1}{4}$ is bigger or smaller than $\frac{2}{7}$.

Arrange the following fractions in ascending order:

$\frac{3}{4}$, $\frac{7}{10}$, $\frac{1}{2}$, $\frac{4}{5}$

$$\frac{3}{4} = \frac{15}{20}$$

$$\frac{7}{10} = \frac{14}{20} \quad \text{(20 divides by 4, 10, 2 and 5)}$$

$$\frac{1}{2} = \frac{10}{20}$$

$$\frac{4}{5} = \frac{16}{20}$$

So the ascending order is $\frac{1}{2}$, $\frac{7}{10}$, $\frac{3}{4}$, $\frac{4}{5}$.

37	$\frac{2}{3}$, $\frac{1}{2}$, $\frac{3}{5}$, $\frac{7}{30}$	**40**	$\frac{2}{5}$, $\frac{3}{8}$, $\frac{17}{20}$, $\frac{1}{2}$, $\frac{7}{10}$
38	$\frac{13}{20}$, $\frac{3}{4}$, $\frac{4}{10}$, $\frac{5}{8}$	**41**	$\frac{5}{7}$, $\frac{11}{14}$, $\frac{3}{4}$, $\frac{17}{28}$, $\frac{1}{2}$
39	$\frac{1}{3}$, $\frac{5}{6}$, $\frac{1}{2}$, $\frac{7}{12}$	**42**	$\frac{7}{10}$, $\frac{2}{5}$, $\frac{3}{5}$, $\frac{14}{25}$, $\frac{1}{2}$

You need to change all four fractions into equivalent fractions with the same denominator, then you can see which is the smallest, which is the next smallest, and so on.

Arrange the following fractions in descending order:

43	$\frac{5}{6}$, $\frac{1}{2}$, $\frac{7}{9}$, $\frac{11}{18}$, $\frac{2}{3}$	45	$\frac{7}{12}$, $\frac{1}{6}$, $\frac{2}{3}$, $\frac{17}{24}$, $\frac{3}{4}$	**47**	$\frac{7}{16}$, $\frac{1}{2}$, $\frac{5}{8}$, $\frac{19}{32}$, $\frac{3}{4}$
44	$\frac{13}{20}$, $\frac{3}{5}$, $\frac{1}{2}$, $\frac{3}{4}$, $\frac{7}{10}$	**46**	$\frac{7}{10}$, $\frac{11}{15}$, $\frac{2}{3}$, $\frac{23}{30}$, $\frac{4}{5}$	**48**	$\frac{4}{5}$, $\frac{7}{12}$, $\frac{5}{6}$, $\frac{1}{2}$, $\frac{3}{4}$

Simplifying fractions

Think of the way you find equivalent fractions.

For example $\qquad \dfrac{2}{5} = \dfrac{2 \times 7}{5 \times 7} = \dfrac{14}{35}$

So, $\qquad \dfrac{14}{35} = \dfrac{\cancel{7} \times 2}{\cancel{7} \times 5} = \dfrac{2}{5}$

In the middle step, 7 is a factor of both the numerator and the denominator and it is called a common factor. To get the final value of $\frac{2}{5}$ we have 'crossed out' the common factor. This is called cancelling. We have divided the top and the bottom numbers by 7. This *simplifies* the fraction.

When all the simplifying is finished we say that the fraction is in its *lowest terms*.

Any fraction whose numerator and denominator have a common factor (perhaps more than one) can be simplified in this way.

For example $\qquad \dfrac{24}{27} = \dfrac{3 \times 8}{3 \times 9} = \dfrac{8}{9}$

A quicker way to write this down is to divide the numerator and the denominator mentally by the common factor, crossing them out and writing the new numbers beside them (it is a good idea to write the new numbers smaller so that you can see that you have simplified the fraction), i.e.

$\dfrac{\overset{8}{\cancel{24}}}{\underset{9}{\cancel{27}}} = \dfrac{8}{9}$

Shortcuts

Find the difference between the numerator and the denominator, and then divide each by the difference.

For example $\qquad$ reduce $\dfrac{112}{119}$ to its lowest terms.

Difference between numerator and denominator $= 119 - 112 = 7$.

$\qquad\qquad 112 \div 7 = 16 \qquad\qquad 119 \div 7 = 17$

$\qquad \therefore \quad \dfrac{112}{119} = \dfrac{16}{17}$

If this does not work, divide by factors of the difference.

For example $\qquad$ reduce $\dfrac{117}{135}$

Difference between numerator and denominator $= 135 - 117 = 18$.
18 is not a common factor.
Try the largest factor of 18, i.e. 9.

$\qquad\qquad 135 \div 9 = 15 \qquad\qquad 117 \div 9 = 13$

$\qquad \therefore \quad \dfrac{117}{135} = \dfrac{13}{15}$

Exercise 5e

Simplify the following fractions:

$$\frac{66}{176} = \frac{3}{8}$$

(We divided top and bottom by 2 and then by 11.)

1	$\frac{2}{6}$	13	$\frac{14}{70}$	25	$\frac{80}{100}$
2	$\frac{30}{50}$	14	$\frac{24}{60}$	26	$\frac{48}{84}$
3	$\frac{3}{9}$	15	$\frac{16}{56}$		
4	$\frac{6}{12}$	16	$\frac{10}{30}$		
5	$\frac{9}{27}$	17	$\frac{36}{72}$		
6	$\frac{4}{8}$	18	$\frac{15}{75}$	27	$\frac{54}{162}$
7	$\frac{5}{15}$	19	$\frac{60}{100}$	28	$\frac{54}{66}$
8	$\frac{12}{18}$	20	$\frac{36}{90}$	29	$\frac{27}{36}$
9	$\frac{10}{20}$	21	$\frac{70}{126}$	30	$\frac{800}{1000}$
10	$\frac{8}{32}$	22	$\frac{49}{77}$		
11	$\frac{8}{28}$	23	$\frac{99}{132}$		
12	$\frac{27}{90}$	24	$\frac{33}{121}$		

When you cancel, check that you have cancelled ALL possible common factors. For example, if you cancel $\frac{6}{12}$ by 2, you are left with $\frac{3}{6}$; this has a common factor of 3 so will simplify further.

Adding fractions

Suppose there is a bowl of oranges and papayas. First you take three oranges and then two more oranges. You then have five oranges; we can add the 3 and the 2 together because they are the same kind of fruit. But three oranges and two papayas cannot be added together because they are different kinds of fruit.

For fractions it is the denominator that tells us the kind of fraction, so we can add fractions together if they have the same denominator but not while their denominators are different.

Exercise 5f

Add the fractions given in questions **1** to **24**, simplifying the answers
where you can.

$\dfrac{9}{22} + \dfrac{5}{22}$

$\dfrac{9}{22} + \dfrac{5}{22} = \dfrac{9+5}{22}$

$= \dfrac{14^{\,7}}{22_{\,11}}$ When you have added the numerators, always check to
see if the fraction will simplify.

$= \dfrac{7}{11}$

1 $\dfrac{1}{4} + \dfrac{2}{4}$ **7** $\dfrac{2}{5} + \dfrac{1}{5}$ **13** $\dfrac{2}{7} + \dfrac{4}{7}$ **19** $\dfrac{3}{20} + \dfrac{7}{20}$

2 $\dfrac{1}{8} + \dfrac{3}{8}$ **8** $\dfrac{3}{10} + \dfrac{1}{10}$ **14** $\dfrac{4}{17} + \dfrac{5}{17}$ **20** $\dfrac{21}{100} + \dfrac{19}{100}$

3 $\dfrac{3}{11} + \dfrac{2}{11}$ **9** $\dfrac{2}{21} + \dfrac{9}{21}$ **15** $\dfrac{3}{14} + \dfrac{4}{14}$ **21** $\dfrac{4}{11} + \dfrac{2}{11}$

4 $\dfrac{3}{13} + \dfrac{7}{13}$ **10** $\dfrac{7}{30} + \dfrac{8}{30}$ **16** $\dfrac{8}{30} + \dfrac{19}{30}$ **22** $\dfrac{14}{23} + \dfrac{1}{23}$

5 $\dfrac{11}{23} + \dfrac{8}{23}$ **11** $\dfrac{6}{13} + \dfrac{5}{13}$ **17** $\dfrac{5}{16} + \dfrac{7}{16}$ **23** $\dfrac{11}{18} + \dfrac{5}{18}$

6 $\dfrac{1}{7} + \dfrac{2}{7}$ **12** $\dfrac{1}{10} + \dfrac{7}{10}$ **18** $\dfrac{8}{19} + \dfrac{3}{19}$ **24** $\dfrac{7}{15} + \dfrac{3}{15}$

We can add more than two fractions in the same way.
Add the fractions given in questions **25** to **34**:

25 $\dfrac{2}{15} + \dfrac{4}{15} + \dfrac{6}{15}$ **29** $\dfrac{2}{51} + \dfrac{4}{51} + \dfrac{6}{51} + \dfrac{8}{51} + \dfrac{7}{51}$ **33** $\dfrac{3}{100} + \dfrac{14}{100} + \dfrac{31}{100} + \dfrac{2}{100}$

26 $\dfrac{8}{100} + \dfrac{21}{100} + \dfrac{11}{100}$ **30** $\dfrac{3}{19} + \dfrac{2}{19} + \dfrac{7}{19}$ **34** $\dfrac{3}{99} + \dfrac{11}{99} + \dfrac{4}{99} + \dfrac{7}{99}$

27 $\dfrac{3}{31} + \dfrac{2}{31} + \dfrac{7}{31} + \dfrac{11}{31}$ **31** $\dfrac{7}{60} + \dfrac{8}{60} + \dfrac{11}{60}$

28 $\dfrac{1}{14} + \dfrac{3}{14} + \dfrac{5}{14} + \dfrac{2}{14}$ **32** $\dfrac{4}{45} + \dfrac{11}{45} + \dfrac{8}{45} + \dfrac{2}{45}$

Fractions with different denominators

To add fractions with different denominators we must first change the
fractions into equivalent fractions with the same denominator. This new
denominator must be a number that both original denominators divide into.
For instance, if we want to add $\dfrac{2}{5}$ and $\dfrac{3}{7}$ we choose 35 for our new denominator
because 35 can be divided by both 5 and 7:

$$\frac{2}{5} = \frac{14}{35}$$

$$\frac{3}{7} = \frac{15}{35}$$

So, $\dfrac{2}{5} + \dfrac{3}{7} = \dfrac{14}{35} + \dfrac{15}{35} = \dfrac{29}{35}$

Exercise 5g

Find $\dfrac{2}{7} + \dfrac{3}{8}$

You need to change $\dfrac{2}{7}$ and $\dfrac{3}{8}$ into equivalent fractions with the same denominator.

7 and 8 both divide into 56 so choose 56 as the new denominator.

$$\frac{2}{7} + \frac{3}{8} = \frac{16}{56} + \frac{21}{56} = \frac{37}{56}$$

Find:

1	$\dfrac{2}{3} + \dfrac{1}{5}$	**4**	$\dfrac{2}{5} + \dfrac{3}{7}$	**7**	$\dfrac{3}{7} + \dfrac{1}{6}$	**10**	$\dfrac{5}{6} + \dfrac{1}{7}$
2	$\dfrac{1}{5} + \dfrac{3}{8}$	**5**	$\dfrac{3}{10} + \dfrac{2}{3}$	**8**	$\dfrac{2}{3} + \dfrac{2}{7}$	**11**	$\dfrac{3}{11} + \dfrac{5}{9}$
3	$\dfrac{1}{5} + \dfrac{1}{6}$	**6**	$\dfrac{4}{7} + \dfrac{1}{8}$	**9**	$\dfrac{1}{6} + \dfrac{2}{7}$	**12**	$\dfrac{2}{9} + \dfrac{3}{10}$

The new denominator, which is called the *common denominator*, is not always as big as you might first think. For instance, if we want to add $\dfrac{3}{4}$ and $\dfrac{1}{12}$, the common denominator is 12 because it divides by both 4 and 12.

Find $\dfrac{3}{4} + \dfrac{1}{12}$

$$\frac{3}{4} + \frac{1}{12} = \frac{9}{12} + \frac{1}{12}$$
$$= \frac{10}{12}^{5}_{6}$$
$$= \frac{5}{6}$$

13	$\dfrac{2}{5} + \dfrac{3}{10}$	**16**	$\dfrac{3}{10} + \dfrac{3}{100}$	**19**	$\dfrac{2}{3} + \dfrac{2}{9}$	**22**	$\dfrac{4}{11} + \dfrac{5}{22}$
14	$\dfrac{3}{8} + \dfrac{7}{16}$	**17**	$\dfrac{1}{4} + \dfrac{7}{10}$	**20**	$\dfrac{4}{9} + \dfrac{5}{18}$	**23**	$\dfrac{2}{5} + \dfrac{7}{15}$
15	$\dfrac{3}{7} + \dfrac{8}{21}$	**18**	$\dfrac{1}{4} + \dfrac{3}{8}$	**21**	$\dfrac{1}{20} + \dfrac{3}{5}$	**24**	$\dfrac{7}{12} + \dfrac{1}{6}$

More than two fractions can be added in this way. The common denominator must be divisible by *all* of the original denominators.

Find $\dfrac{1}{8} + \dfrac{1}{2} + \dfrac{1}{3}$

(8, 2 and 3 all divide into 24)

$$\frac{1}{8} + \frac{1}{2} + \frac{1}{3} = \frac{3}{24} + \frac{12}{24} + \frac{8}{24}$$
$$= \frac{3 + 12 + 8}{24}$$
$$= \frac{23}{24}$$

25 $\frac{1}{5}+\frac{1}{4}+\frac{1}{2}$

28 $\frac{5}{12}+\frac{1}{6}+\frac{1}{3}$

31 $\frac{1}{2}+\frac{3}{8}+\frac{1}{10}$

34 $\frac{2}{9}+\frac{2}{3}+\frac{1}{18}$

26 $\frac{1}{8}+\frac{1}{4}+\frac{1}{3}$

29 $\frac{1}{7}+\frac{3}{14}+\frac{1}{2}$

32 $\frac{1}{3}+\frac{2}{9}+\frac{1}{6}$

35 $\frac{2}{15}+\frac{1}{10}+\frac{2}{5}$

27 $\frac{3}{10}+\frac{2}{5}+\frac{1}{4}$

30 $\frac{1}{3}+\frac{1}{6}+\frac{1}{2}$

33 $\frac{7}{20}+\frac{3}{10}+\frac{1}{5}$

36 $\frac{1}{4}+\frac{1}{12}+\frac{1}{3}$

Subtracting fractions

Exactly the same method is used for subtracting fractions as for adding them.
To work out the value of $\frac{7}{8}-\frac{3}{8}$ we notice that the denominators are the same, so

$$\frac{7}{8}-\frac{3}{8}=\frac{7-3}{8}$$
$$=\frac{4}{8}$$
$$=\frac{1}{2}$$

Exercise 5h

Find $\frac{7}{9}-\frac{1}{4}$

(The denominators are not the same so we use equivalent fractions
with denominator 36.)

$$\frac{7}{9}-\frac{1}{4}=\frac{28}{36}-\frac{9}{36}$$
$$=\frac{28-9}{36}$$
$$=\frac{19}{36} \text{ (This will not simplify.)}$$

Find:

1 $\frac{8}{9}-\frac{2}{9}$

7 $\frac{8}{13}-\frac{3}{13}$

13 $\frac{8}{11}-\frac{2}{5}$

19 $\frac{15}{16}-\frac{3}{4}$

2 $\frac{7}{10}-\frac{2}{10}$

8 $\frac{19}{20}-\frac{7}{20}$

14 $\frac{7}{9}-\frac{2}{3}$

20 $\frac{7}{15}-\frac{1}{5}$

3 $\frac{6}{17}-\frac{1}{17}$

9 $\frac{2}{3}-\frac{3}{7}$

15 $\frac{8}{13}-\frac{1}{2}$

21 $\frac{3}{4}-\frac{5}{8}$

4 $\frac{3}{4}-\frac{1}{5}$

10 $\frac{4}{7}-\frac{1}{3}$

16 $\frac{11}{12}-\frac{5}{6}$

22 $\frac{7}{12}-\frac{1}{3}$

5 $\frac{9}{10}-\frac{1}{2}$

11 $\frac{11}{15}-\frac{4}{15}$

17 $\frac{19}{100}-\frac{1}{10}$

23 $\frac{13}{18}-\frac{5}{9}$

6 $\frac{5}{7}-\frac{2}{7}$

12 $\frac{13}{18}-\frac{7}{18}$

18 $\frac{5}{8}-\frac{2}{7}$

24 $\frac{13}{15}-\frac{3}{5}$

Adding and subtracting fractions

Fractions can be added and subtracted in one problem in a similar way.

For example

$$\frac{7}{9} + \frac{1}{18} - \frac{1}{6} = \frac{14}{18} + \frac{1}{18} - \frac{3}{18}$$

$$= \frac{14 + 1 - 3}{18}$$

$$= \frac{15 - 3}{18}$$

$$= \frac{\overset{2}{\cancel{12}}}{\underset{3}{\cancel{18}}} \quad \text{This will simplify by cancelling by 6.}$$

$$= \frac{2}{3}$$

It is not always possible to work from left to right in order because we have to subtract too much too soon. In this case we can do the adding first. Remember that it is the operation (i.e. add or subtract) *in front* of a number that tells you what to do with that number.

This flow chart shows the process for adding and subtracting fractions.

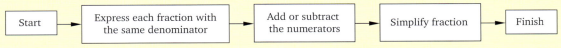

| Start | → | Express each fraction with the same denominator | → | Add or subtract the numerators | → | Simplify fraction | → | Finish |

Exercise 5i

Find $\dfrac{1}{8} - \dfrac{3}{4} + \dfrac{11}{16}$

$$\frac{1}{8} - \frac{3}{4} + \frac{11}{16} = \frac{2}{16} - \frac{12}{16} + \frac{11}{16} = \frac{2}{16} + \frac{11}{16} - \frac{12}{16}$$

$$= \frac{2 + 11 - 12}{16} = \frac{13 - 12}{16}$$

$$= \frac{1}{16}$$

Find:

1 $\dfrac{3}{4} + \dfrac{1}{2} - \dfrac{7}{8}$

2 $\dfrac{6}{7} - \dfrac{9}{14} + \dfrac{1}{2}$

3 $\dfrac{3}{8} + \dfrac{7}{16} - \dfrac{3}{4}$

4 $\dfrac{11}{12} + \dfrac{1}{6} - \dfrac{2}{3}$

5 $\dfrac{3}{5} + \dfrac{3}{25} - \dfrac{27}{50}$

6 $\dfrac{2}{3} + \dfrac{1}{6} - \dfrac{5}{12}$

7 $\dfrac{4}{5} - \dfrac{7}{10} + \dfrac{1}{2}$

8 $\dfrac{7}{9} - \dfrac{2}{3} + \dfrac{5}{6}$

9 $\dfrac{7}{10} - \dfrac{41}{100} + \dfrac{1}{20}$

10 $\dfrac{5}{8} - \dfrac{21}{40} + \dfrac{2}{5}$

Remember that the operation (+ or −) in front of a number tells you what to do with that number only.

11 $\frac{7}{12} - \frac{1}{6} + \frac{1}{3}$

12 $\frac{2}{3} - \frac{7}{18} + \frac{2}{9}$

13 $\frac{2}{9} - \frac{1}{3} + \frac{1}{6}$

14 $\frac{1}{6} - \frac{2}{3} + \frac{7}{12}$

15 $\frac{2}{5} - \frac{1}{2} + \frac{3}{10}$

16 $\frac{1}{8} - \frac{13}{16} + \frac{3}{4}$

17 $\frac{1}{6} - \frac{5}{18} + \frac{1}{3}$

18 $\frac{1}{5} - \frac{7}{10} + \frac{17}{20}$

19 $\frac{1}{4} - \frac{5}{8} + \frac{1}{2}$

20 $\frac{2}{3} - \frac{5}{6} + \frac{1}{2}$

21 $\frac{3}{10} - \frac{61}{100} + \frac{1}{2}$

22 $\frac{1}{8} - \frac{7}{24} + \frac{5}{12}$

23 $\frac{1}{3} - \frac{5}{18} + \frac{2}{9}$

24 $\frac{3}{10} + \frac{2}{15} - \frac{2}{5}$

Remember that you can do the addition before you do the subtraction. For example, to find $\frac{4-6+3}{18}$, you can add 4 and 3 before taking 6 away.

! Investigation

Can you determine how many handshakes there would be if each member in a class of thirty decides to shake hands with every other member of the class?

Consider the following table and fill in the blank spaces.

No. of people, n	Number of handshakes, h	Difference in h's
1	0	0
2	1	1
3	3	2
4	6	3
5	10	4
6		
7		
8		
9		
*		
*		
*		
20	190	19

How many handshakes will there be for a class of 30?

How many handshakes will there be for 100 persons?

Problems

Exercise 5j

In a class of school children, $\frac{1}{3}$ of the children come to school by bus, $\frac{1}{4}$ come to school on bicycles and the rest walk to school. What fraction of the children ride to school? What fraction do not use a bus?

Riding to school means coming by bus or on a bicycle.

The fraction who ride to school on bicycle or bus $= \frac{1}{3} + \frac{1}{4}$

$$= \frac{4+3}{12}$$

$$= \frac{7}{12}$$

Therefore $\frac{7}{12}$ of the children ride to school.

The complete class of children is a whole unit, i.e. 1.

The fraction of children who do not use a bus is found by taking the bus users from the complete class, i.e.

$$\frac{1}{1} - \frac{1}{3} = \frac{3-1}{3} = \frac{2}{3}$$

Remember to read the questions carefully. Read each one several times if necessary to make sure that you understand what you are being asked to find.

1 A girl spends $\frac{1}{5}$ of her pocket money on sweets and $\frac{2}{3}$ on records. What fraction has she spent? What fraction has she left?

2 A group of friends went to a hamburger bar. $\frac{2}{5}$ of them bought a hamburger, $\frac{1}{3}$ of them just bought fries. The rest bought cola. What fraction of the group bought food? What fraction bought a drink?

3 At a calypso final, $\frac{2}{3}$ of the groups were all male, $\frac{1}{4}$ of the groups included one girl and the rest included more than one girl. What fraction of the groups
  a were not all male   **b** included more than one girl?

4 At a youth club, $\frac{1}{2}$ of the meetings are for playing table tennis only, $\frac{1}{8}$ of the meetings are discussions only and the rest are record sessions only. What fraction of the meetings are
  a record sessions only   **b** not for discussions?

5 At a school, $\frac{1}{8}$ of the time is spent in mathematics classes, $\frac{3}{20}$ of the time in English classes and $\frac{1}{20}$ on games. What fraction of the time is spent on
  a English and maths together   **b** all lessons except games
  c maths and games?

? **Puzzle**

An old lady went to market with a basket of eggs. To the first customer she sold half of what she had plus half an egg. The second customer bought half of what remained plus half an egg and the third customer bought half of what now remained and half an egg. That left the lady with thirty-six eggs. There were no broken eggs at any time. How many eggs did she have in her basket to start?

Mixed numbers and improper fractions

Most of the fractions we have met so far have been less than a whole unit. These are called *proper* fractions. But we often have more than a whole unit. Suppose, for instance, that we have one and a half bars of chocolate:

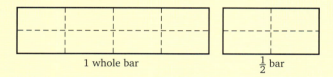

1 whole bar $\frac{1}{2}$ bar

We have $1\frac{1}{2}$ bars, and $1\frac{1}{2}$ is called a *mixed number*.

Another way of describing the amount of chocolate is to say that we have three half bars.

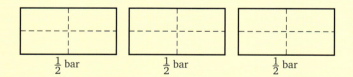

$\frac{1}{2}$ bar $\frac{1}{2}$ bar $\frac{1}{2}$ bar

We have $\frac{3}{2}$ bars and $\frac{3}{2}$ is called an *improper* fraction because the numerator is bigger than the denominator.

But the amount of chocolate in the two examples is the same,

so $\frac{3}{2} = 1\frac{1}{2}$

Improper fractions can be changed into mixed numbers by finding out how many whole units there are. For instance, to change $\frac{8}{3}$ into a mixed number we look for the biggest number below 8 that divides by 3, i.e. 6.

Then $\frac{8}{3} = \frac{6+2}{3} = \frac{6}{3} + \frac{2}{3} = 2 + \frac{2}{3} = 2\frac{2}{3}$

Exercise 5k

In questions **1** to **20** change the improper fractions into mixed numbers:

1 $\frac{9}{4}$ **8** $\frac{41}{8}$ **15** $\frac{87}{11}$ **18** $\frac{67}{5}$

2 $\frac{19}{4}$ **9** $\frac{127}{5}$ **16** $\frac{77}{6}$ **19** $\frac{73}{3}$

3 $\frac{37}{6}$ **10** $\frac{114}{11}$ **17** $\frac{41}{3}$ **20** $\frac{49}{10}$

4 $\frac{53}{10}$ **11** $\frac{109}{8}$

5 $\frac{88}{9}$ **12** $\frac{83}{7}$

6 $\frac{7}{2}$ **13** $\frac{121}{9}$

7 $\frac{27}{4}$ **14** $\frac{91}{6}$

Find the number of 4s in 9: this gives the units. The remainder is the numbers of quarters.

We can also change mixed numbers into improper fractions. For instance, in $2\frac{4}{5}$ we have two whole units and $\frac{4}{5}$. In each whole unit there are five fifths, so in $2\frac{4}{5}$ we have ten fifths and four fifths, i.e.

$$2\frac{4}{5} = \frac{10}{5} + \frac{4}{5} = \frac{14}{5}$$

Exercise 5l

In questions **1** to **20** change the mixed numbers into improper fractions.

So, $3\frac{1}{7}$

$$3\frac{1}{7} = 3 + \frac{1}{7}$$
$$= \frac{21}{7} + \frac{1}{7}$$
$$= \frac{22}{7}$$

1 $4\frac{1}{3}$ **6** $6\frac{3}{5}$ **11** $7\frac{2}{5}$ **16** $10\frac{3}{7}$

2 $8\frac{1}{4}$ **7** $2\frac{6}{7}$ **12** $2\frac{4}{9}$ **17** $1\frac{9}{10}$

3 $1\frac{7}{10}$ **8** $4\frac{1}{6}$ **13** $3\frac{4}{5}$ **18** $6\frac{2}{3}$

4 $10\frac{8}{9}$ **9** $3\frac{2}{3}$ **14** $4\frac{7}{9}$ **19** $7\frac{3}{8}$

5 $8\frac{1}{7}$ **10** $5\frac{1}{2}$ **15** $8\frac{3}{4}$ **20** $10\frac{1}{10}$

Fractions as a result of division

$15 \div 4$ means 'how many fours are there in 15?'

There are 3 fours in 15 with 3 left over, so $15 \div 4 = 3$, remainder 3.

Note that the remainder, 3, is $\frac{3}{4}$ of 4. Thus we can say that there are $3\frac{3}{4}$ fours in 15.

i.e. $\qquad\qquad 15 \div 4 = 3\frac{3}{4}$

But $\qquad\qquad\quad \frac{15}{4} = 3\frac{3}{4}$

Therefore $15 \div 4$ and $\frac{15}{4}$ mean the same thing.

Exercise 5m

Find $27 \div 8$

$$27 \div 8 = \frac{27}{8}$$

There are 3 eights in 27 with 3 left over: so there are 3 units and 3 eighths.

$$= 3\frac{3}{8}$$

Calculate the following divisions, giving your answers as mixed numbers:

1	$36 \div 7$	**4**	$20 \div 8$	**7**	$41 \div 3$	**10**	$107 \div 10$
2	$59 \div 6$	**5**	$82 \div 5$	**8**	$64 \div 9$	**11**	$37 \div 5$
3	$52 \div 11$	**6**	$29 \div 4$	**9**	$98 \div 12$	**12**	$52 \div 8$

Adding mixed numbers

If we want to find the value of $2\frac{1}{3} + 3\frac{1}{4}$ we add the whole numbers and then the fractions, i.e.

$$2\frac{1}{3} + 3\frac{1}{4} = 2 + 3 + \frac{1}{3} + \frac{1}{4}$$
$$= 5 + \frac{4+3}{12}$$
$$= 5 + \frac{7}{12}$$
$$= 5\frac{7}{12}$$

Sometimes there is an extra step in the calculation.

For example
$$3\tfrac{1}{2} + 2\tfrac{3}{8} + 5\tfrac{1}{4} = 3 + 2 + 5 + \tfrac{1}{2} + \tfrac{3}{8} + \tfrac{1}{4}$$
$$= 10 + \frac{4+3+2}{8}$$
$$= 10 + \tfrac{9}{8}$$

But $\tfrac{9}{8}$ is an improper fraction, so we change it into a mixed number

i.e.
$$3\tfrac{1}{2} + 2\tfrac{3}{8} + 5\tfrac{1}{4} = 10 + \frac{8+1}{8}$$
$$= 10 + 1 + \tfrac{1}{8}$$
$$= 11\tfrac{1}{8}$$

Exercise 5n

Find:

1 $2\tfrac{1}{4} + 3\tfrac{1}{2}$

2 $1\tfrac{1}{2} + 2\tfrac{1}{3}$

3 $4\tfrac{1}{5} + 1\tfrac{3}{8}$

4 $5\tfrac{1}{9} + 4\tfrac{1}{3}$

5 $3\tfrac{1}{4} + 2\tfrac{5}{9}$

6 $1\tfrac{1}{3} + 2\tfrac{5}{6}$

7 $3\tfrac{1}{4} + 1\tfrac{1}{5}$

8 $2\tfrac{1}{7} + 1\tfrac{1}{14}$

9 $6\tfrac{3}{10} + 1\tfrac{2}{5}$

10 $8\tfrac{1}{7} + 5\tfrac{2}{3}$

11 $7\tfrac{3}{8} + 3\tfrac{7}{16}$

12 $1\tfrac{3}{4} + 4\tfrac{7}{12}$

13 $3\tfrac{5}{7} + 7\tfrac{1}{2}$

14 $6\tfrac{1}{2} + 1\tfrac{9}{16}$

15 $8\tfrac{7}{8} + 3\tfrac{3}{16}$

16 $2\tfrac{7}{10} + 9\tfrac{1}{5}$

17 $5\tfrac{7}{10} + 2\tfrac{3}{5}$

18 $9\tfrac{2}{3} + 8\tfrac{5}{6}$

19 $2\tfrac{4}{5} + 7\tfrac{3}{10}$

20 $6\tfrac{3}{10} + 4\tfrac{4}{5}$

21 $1\tfrac{1}{4} + 3\tfrac{2}{3} + 6\tfrac{7}{12}$

22 $5\tfrac{1}{7} + 4\tfrac{1}{2} + 7\tfrac{11}{14}$

23 $3\tfrac{3}{4} + 5\tfrac{1}{8} + 8\tfrac{5}{16}$

24 $10\tfrac{2}{3} + 3\tfrac{1}{6} + 7\tfrac{2}{9}$

25 $4\tfrac{4}{5} + 9\tfrac{4}{15} + 1\tfrac{1}{3}$

26 $4\tfrac{3}{5} + 8\tfrac{7}{10} + 2\tfrac{1}{2}$

27 $3\tfrac{7}{10} + 9\tfrac{21}{100} + 1\tfrac{3}{5}$

28 $4\tfrac{1}{4} + 7\tfrac{1}{8} + 6\tfrac{1}{32}$

29 $1\tfrac{5}{7} + 11\tfrac{1}{2} + 9\tfrac{1}{14}$

30 $10\tfrac{7}{9} + 6\tfrac{1}{3} + 5\tfrac{7}{18}$

 Remember that $2\tfrac{1}{4}$ means $2 + \tfrac{1}{4}$ and that $3\tfrac{1}{2}$ means $3 + \tfrac{1}{2}$. Add the units first then add the fractions.

 Remember to check the fraction in your answer: if it is improper change it to a mixed number, then add the units.

Subtracting mixed numbers

If we want to find the value of $5\frac{3}{4} - 2\frac{2}{5}$ we can use the same method as for adding:

$$5\frac{3}{4} - 2\frac{2}{5} = 5 - 2 + \frac{3}{4} - \frac{2}{5}$$
$$= 3 + \frac{15 - 8}{20}$$
$$= 3 + \frac{7}{20}$$
$$= 3\frac{7}{20}$$

But when we find the value of $6\frac{1}{4} - 2\frac{4}{5}$ we get

$$6\frac{1}{4} - 2\frac{4}{5} = 6 - 2 + \frac{1}{4} - \frac{4}{5}$$
$$= 4 + \frac{1}{4} - \frac{4}{5}$$

This time it is not so easy to deal with the fractions because $\frac{4}{5}$ is bigger than $\frac{1}{4}$. So we take one of the whole units and change it into a fraction, giving

$$3 + 1 + \frac{1}{4} - \frac{4}{5}$$
$$= 3 + \frac{20 + 5 - 16}{20}$$
$$= 3 + \frac{9}{20}$$
$$= 3\frac{9}{20}$$

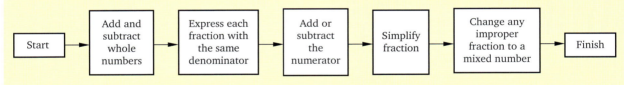

| Start | → | Add and subtract whole numbers | → | Express each fraction with the same denominator | → | Add or subtract the numerator | → | Simplify fraction | → | Change any improper fraction to a mixed number | → | Finish |

Exercise 5p

Find:

1 $2\frac{3}{4} - 1\frac{1}{8}$

2 $3\frac{2}{3} - 1\frac{4}{5}$

3 $1\frac{5}{6} - \frac{2}{3}$

4 $3\frac{1}{4} - 2\frac{1}{2}$

5 $7\frac{3}{4} - 2\frac{1}{3}$

6 $3\frac{5}{6} - 2\frac{1}{3}$

7 $2\frac{6}{7} - 1\frac{1}{2}$

8 $4\frac{1}{2} - 2\frac{1}{5}$

9 $4\frac{4}{5} - 3\frac{1}{10}$

10 $6\frac{5}{7} - 3\frac{2}{5}$

11 $3\frac{1}{3} - 1\frac{1}{5}$

12 $5\frac{3}{4} - 2\frac{1}{2}$

13 $8\frac{4}{5} - 5\frac{1}{2}$

14 $5\frac{7}{9} - 3\frac{5}{7}$

15 $4\frac{5}{8} - 1\frac{1}{3}$

16 $6\frac{3}{4} - 3\frac{6}{7}$

17 $7\frac{1}{2} - 5\frac{3}{4}$

18 $4\frac{3}{5} - 1\frac{1}{4}$

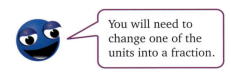

You will need to change one of the units into a fraction.

19 $7\frac{6}{7} - 4\frac{3}{5}$ **24** $5\frac{4}{7} - 3\frac{4}{5}$ **29** $8\frac{2}{3} - 7\frac{8}{9}$ **34** $2\frac{5}{12} - 1\frac{3}{4}$

20 $8\frac{8}{11} - 2\frac{2}{3}$ **25** $3\frac{1}{4} - 1\frac{7}{8}$ **30** $4\frac{1}{6} - 2\frac{2}{3}$ **35** $4\frac{7}{9} - 3\frac{11}{18}$

21 $8\frac{6}{7} - 5\frac{3}{4}$ **26** $5\frac{3}{5} - 2\frac{9}{10}$ **31** $6\frac{2}{3} - 3\frac{5}{6}$ **36** $5\frac{1}{3} - 2\frac{4}{7}$

22 $3\frac{1}{2} - 1\frac{7}{8}$ **27** $9\frac{7}{10} - 6\frac{1}{5}$ **32** $7\frac{3}{4} - 4\frac{7}{8}$

23 $2\frac{1}{2} - 1\frac{3}{4}$ **28** $6\frac{3}{10} - 3\frac{4}{5}$ **33** $9\frac{7}{10} - 5\frac{4}{5}$

(?) Puzzle

Fred is an old man. He lived one-eighth of his life as a boy, one-twelfth as a youth, one-half as a man and has spent 28 years in his old age. How old is Fred now?

Mixed exercises

Exercise 5q

1 Calculate:
  a $\frac{2}{3} + \frac{4}{7}$ **b** $\frac{5}{6} - \frac{3}{8}$ **c** $\frac{3}{8} + \frac{1}{9}$
  d $1\frac{1}{2} + \frac{2}{3}$ **e** $2\frac{1}{4} - 1\frac{1}{3}$

2 Simplify:
  a $\frac{54}{24}$ **b** $3\frac{15}{75}$

3 Write the first quantity as a fraction of the second quantity:
  a 3 days; 1 week   **b** 17 children; 30 children

4 Write the following fractions in ascending size order:
  a $\frac{1}{2}, \frac{7}{10}, \frac{3}{5}, \frac{13}{20}$ **b** $\frac{3}{4}, \frac{7}{12}, \frac{5}{6}, \frac{2}{3}$ **c** $\frac{3}{5}, \frac{7}{10}, \frac{17}{20}, \frac{71}{100}$

5 Write either > or < between the following pairs of fractions, to make true statements:
  a $\frac{5}{12}$ $\frac{7}{16}$ **b** $\frac{3}{8}$ $\frac{7}{24}$ **c** $\frac{13}{22}$ $\frac{19}{33}$

6 A cricket club consists of 7 members who are good batsmen only, 5 who are good bowlers only, 4 all-rounders and some non-players. If there are 22 people in the club, what fraction of them are
  a non-players   **b** good batsmen only.

Exercise 5r

1   Calculate:
    a $\frac{4}{5} - \frac{2}{3}$
    b $\frac{9}{10} + \frac{4}{5}$
    c $\frac{7}{11} - \frac{1}{2}$
    d $2\frac{1}{3} + 4\frac{1}{4}$
    e $3\frac{1}{6} - 2\frac{2}{3}$
    f $5\frac{1}{4} - 2\frac{3}{5}$

2   Simplify:
    a $\frac{84}{96}$
    b $\frac{77}{42}$
    c $\frac{84}{91}$

3   Write the first quantity as a fraction of the second quantity:
    a $13; $100
    b   233 days; 1 leap year

4   Write either < or > between the following pairs of fractions:
    a $\frac{13}{20}$ $\frac{7}{15}$
    b $\frac{5}{9}$ $\frac{11}{18}$
    c $\frac{5}{6}$ $\frac{7}{8}$

5   Write the following fractions in ascending size order:
    a $\frac{7}{20}, \frac{3}{8}, \frac{2}{5}, \frac{3}{10}$
    b $\frac{2}{5}, \frac{7}{15}, \frac{3}{10}, \frac{1}{2}$
    c $\frac{9}{16}, \frac{3}{4}, \frac{5}{8}, \frac{17}{32}$

6   In a class of 28 children, 13 live in houses with gardens, 7 live in houses without gardens and the rest live in flats. What fraction of the children
    a   do not live in houses with gardens
    b   live in flats?

Exercise 5s

1   Calculate:
    a $\frac{6}{7} + \frac{1}{5} - \frac{3}{4}$
    b $\frac{3}{5} + \frac{4}{9} - \frac{2}{3}$
    c $\frac{1}{2} + \frac{7}{8} - 1\frac{1}{4}$
    d $2\frac{1}{4} - 1\frac{1}{3} + 2\frac{1}{6}$
    e $4\frac{1}{5} - 5\frac{1}{2} + 1\frac{3}{10}$
    f $6\frac{1}{3} - 2\frac{4}{5} + 1\frac{7}{15}$

2   Simplify:
    a $1\frac{18}{48}$
    b $2\frac{18}{45}$
    c $\frac{10}{32}$

3   Write either > or < between the following pairs of fractions:
    a $\frac{5}{8}$ $\frac{7}{11}$
    b $\frac{4}{5}$ $\frac{5}{6}$

4   Arrange the following fractions in ascending size order:
    a $\frac{3}{4}, \frac{3}{5}, \frac{1}{2}, \frac{5}{6}$
    b $\frac{1}{2}, \frac{5}{6}, \frac{5}{9}, \frac{2}{3}$

5   Write the first quantity as a fraction of the second quantity:
    a   7 minutes; 1 hour
    b   1200 people; 3600 people
    c $76; $158

6   In a bag of yams there are 6 large good ones, 11 small good ones and 2 rotten ones. What fraction of the yams in the bag are
    a   good ones
    b   not large good ones?

Exercise 5t

1 Calculate:

  a $\dfrac{4}{5}+\dfrac{2}{3}-\dfrac{3}{10}$ b $\dfrac{7}{8}+\dfrac{3}{5}-\dfrac{17}{20}$ c $1\dfrac{1}{3}+\dfrac{5}{6}-2\dfrac{1}{12}$

  d $3\dfrac{1}{5}-2\dfrac{1}{4}+1\dfrac{1}{2}$ e $2\dfrac{1}{3}-3\dfrac{1}{4}+1\dfrac{5}{6}$ f $6\dfrac{3}{4}-4\dfrac{2}{3}+1\dfrac{7}{12}$

2 Simplify:

  a $4\dfrac{12}{32}$ b $\dfrac{30}{240}$ c $\dfrac{108}{42}$

3 Write the first quantity as a fraction of the second quantity:

  a 5 hours; 24 hours

  b 6 seconds; 1 minute

  c 5 months; 1 year

4 Write either > or < between the following pairs of fractions:

  a $\dfrac{7}{10}$ $\dfrac{5}{9}$ b $\dfrac{2}{3}$ $\dfrac{5}{7}$

5 Arrange the following fractions in ascending size order:

  a $\dfrac{5}{11}, \dfrac{1}{2}, \dfrac{13}{22}, \dfrac{23}{44}$ b $\dfrac{2}{3}, \dfrac{5}{9}, \dfrac{3}{4}, \dfrac{7}{12}$

6 A vase of flowers contains 5 pink ones, 3 red ones and 7 white ones. What fraction of the flowers in the vase are

  a red   b not white   c pink?

In this chapter you have seen that...

✔ you can find a fraction equivalent to a given fraction by multiplying the numerator and denominator by the same number

✔ you can simplify fractions by dividing the numerator and denominator by their common factors

✔ you can add and subtract fractions by changing them to equivalent fractions with the same denominator

✔ you can find one quantity as a fraction of another by first making sure that they are in the same units, then by placing the first quantity over the second

✔ $15 \div 4$ and $\dfrac{15}{4}$ mean the same thing

✔ you can compare the size of fractions by changing them to equivalent fractions with the same denominator and then comparing their numerators

✔ as with whole numbers, the order in which you add and subtract fractions does not matter, but it is often convenient to do the addition first.

6 Fractions: multiplication and division

At the end of this chapter you should be able to...

1 Solve word problems involving fractional operations.

2 Perform operations involving multiplication and division of fractions.

3 Express improper fractions as mixed numbers.

You need to know...

✔ how to add and subtract fractions

✔ how to simplify a fraction

✔ how to express a mixed number as an improper fraction and how to express an improper fraction as a mixed number

✔ how to convert a fraction into an equivalent fraction with a different denominator

Key words

cancel, common factor, denominator, expression, improper fraction, mixed number, mixed operation, numerator, product

Multiplying fractions

When fractions are multiplied the result is given by multiplying together the numbers in the numerator and also multiplying together the numbers in the denominator.

For example $\quad \frac{1}{2} \times \frac{1}{3} = \frac{1 \times 1}{2 \times 3}$

$$= \frac{1}{6}$$

If we look at a cake diagram we can see that $\frac{1}{2}$ of $\frac{1}{3}$ of the cake is $\frac{1}{6}$ of the cake.

So $\qquad \frac{1}{2}$ of $\frac{1}{3} = \frac{1}{6}$

and $\qquad \frac{1}{2} \times \frac{1}{3} = \frac{1}{6}$

We see that 'of' means 'multiplied by'.

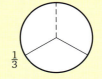 

$\frac{1}{3}$ $\qquad$ $\frac{1}{2}$ of $\frac{1}{3}$

Exercise 6a

Draw cake diagrams to show that:

1 $\frac{1}{2} \times \frac{1}{4} = \frac{1}{8}$ **3** $\frac{1}{2} \times \frac{3}{4} = \frac{3}{8}$ **5** $\frac{1}{3} \times \frac{2}{5} = \frac{2}{15}$

2 $\frac{1}{3} \times \frac{1}{2} = \frac{1}{6}$ **4** $\frac{2}{3} \times \frac{1}{3} = \frac{2}{9}$ **6** $\frac{1}{4} \times \frac{1}{3} = \frac{1}{12}$

Simplifying

Sometimes we can simplify a product by cancelling the common factors.

For example

$$\frac{2}{3} \times \frac{3}{4} = \frac{2 \times 3}{3 \times 4} = \frac{2}{3} \times \frac{3}{4} = \frac{1 \times 1}{1 \times 2}$$

$$= \frac{1}{2}$$

Now you can see that 3 is a common factor of the top and the bottom, so can be cancelled.

The diagram shows that

$\frac{2}{3}$ of $\frac{3}{4} = \frac{1}{2}$

Exercise 6b

Find $\frac{4}{25} \times \frac{15}{16}$

$$\frac{4}{25} \times \frac{15}{16}$$

Cancel 5, which is a common factor of 15 and 25, and 4, which is a common factor of 4 and 16.

$$= \frac{1 \times 3}{5 \times 4} = \frac{3}{20}$$

Find:

1 $\frac{3}{4} \times \frac{1}{2}$ **7** $\frac{3}{7} \times \frac{2}{5}$ **13** $\frac{7}{8} \times \frac{4}{21}$ **19** $\frac{4}{5} \times \frac{15}{16}$

2 $\frac{2}{3} \times \frac{5}{7}$ **8** $\frac{2}{5} \times \frac{3}{5}$ **14** $\frac{3}{4} \times \frac{16}{21}$ **20** $\frac{10}{11} \times \frac{33}{35}$

3 $\frac{2}{5} \times \frac{1}{3}$ **9** $\frac{5}{6} \times \frac{1}{4}$ **15** $\frac{21}{22} \times \frac{11}{27}$ **21** $\frac{4}{15} \times \frac{25}{64}$

4 $\frac{1}{2} \times \frac{7}{8}$ **10** $\frac{2}{3} \times \frac{7}{9}$ **16** $\frac{8}{9} \times \frac{33}{44}$ **22** $\frac{2}{3} \times \frac{33}{40}$

5 $\frac{3}{4} \times \frac{4}{7}$ **11** $\frac{3}{4} \times \frac{1}{5}$ **17** $\frac{7}{9} \times \frac{3}{21}$ **23** $\frac{3}{7} \times \frac{28}{33}$

6 $\frac{4}{9} \times \frac{1}{7}$ **12** $\frac{1}{7} \times \frac{3}{5}$ **18** $\frac{3}{4} \times \frac{5}{7}$ **24** $\frac{48}{55} \times \frac{5}{12}$

Find:

25 $\dfrac{3}{7} \times \dfrac{5}{9} \times \dfrac{14}{15}$

26 $\dfrac{11}{21} \times \dfrac{30}{31} \times \dfrac{7}{55}$

27 $\dfrac{15}{16} \times \dfrac{8}{9} \times \dfrac{4}{5}$

28 $\dfrac{5}{6} \times \dfrac{8}{25} \times \dfrac{3}{4}$

29 $\dfrac{3}{10} \times \dfrac{5}{9} \times \dfrac{6}{7}$

30 $\dfrac{5}{7} \times \dfrac{3}{8} \times \dfrac{21}{30}$

31 $\dfrac{1}{2} \times \dfrac{7}{12} \times \dfrac{18}{35}$

32 $\dfrac{7}{11} \times \dfrac{8}{9} \times \dfrac{33}{28}$

33 $\dfrac{6}{5} \times \dfrac{4}{3} \times \dfrac{10}{4}$

34 $\dfrac{9}{8} \times \dfrac{1}{3} \times \dfrac{4}{27}$

35 $\dfrac{7}{16} \times \dfrac{9}{11} \times \dfrac{8}{21}$

36 $\dfrac{5}{14} \times \dfrac{21}{25} \times \dfrac{5}{9}$

> Express as $\dfrac{3 \times 5 \times 14}{7 \times 9 \times 15}$ then look for common factors in the top and bottom.

? Puzzle

What is one-half of two-thirds of three-quarters of four-fifths of five-sixths of six-sevenths of seven-eighths of eight-ninths of nine-tenths of 500?

Multiplying mixed numbers

Suppose that we want to find the value of $2\frac{1}{3} \times \dfrac{5}{21} \times 1\frac{1}{5}$.

We cannot multiply mixed numbers together unless we change them into improper fractions first. So we change $2\frac{1}{3}$ into $\dfrac{7}{3}$ and we change $1\frac{1}{5}$ into $\dfrac{6}{5}$

Then we can use the same method as before, e.g. $2\frac{1}{3} \times \dfrac{5}{21} \times 1\frac{1}{5} = \dfrac{7 \times 5 \times 6}{3 \times 21 \times 5}$

Now look for common factors to cancel.

Start → Change mixed numbers to improper fractions → Cancel common factors in the numerators and denominators → Multiply the numerators and multiply the denominators → Finish

Exercise 6c

Find $2\frac{1}{3} \times \dfrac{5}{21} \times 1\frac{1}{5}$

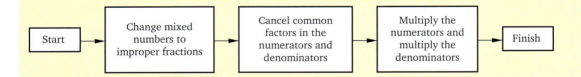

$$2\frac{1}{3} \times \frac{5}{21} \times 1\frac{1}{5} = \frac{7}{3} \times \frac{5}{21} \times \frac{6}{5}$$

$$= \frac{7 \times 5 \times 6}{3 \times 21 \times 5}$$

$$= \frac{2}{3}$$

Find:

1   $1\frac{1}{2} \times \frac{2}{5}$

2   $2\frac{1}{2} \times \frac{4}{5}$

3   $3\frac{1}{4} \times \frac{3}{13}$

4   $4\frac{2}{3} \times 2\frac{2}{5}$

5   $2\frac{1}{5} \times \frac{5}{22}$

6   $1\frac{1}{4} \times \frac{2}{5}$

7   $2\frac{1}{3} \times \frac{3}{8}$

8   $\frac{10}{11} \times 2\frac{1}{5}$

9   $3\frac{1}{2} \times 4\frac{2}{3}$

10 $4\frac{1}{4} \times \frac{4}{21}$

11 $5\frac{1}{4} \times 2\frac{2}{3}$

12 $3\frac{5}{7} \times 1\frac{1}{13}$

13 $8\frac{1}{3} \times 3\frac{3}{5}$

14 $2\frac{1}{10} \times 7\frac{6}{7}$

15 $6\frac{3}{10} \times 1\frac{4}{21}$

16 $4\frac{2}{7} \times 2\frac{1}{10}$

17 $6\frac{1}{4} \times 1\frac{3}{5}$

18 $5\frac{1}{2} \times 1\frac{9}{11}$

19 $8\frac{3}{4} \times 2\frac{2}{7}$

20 $16\frac{1}{2} \times 3\frac{7}{11}$

21 $6\frac{2}{5} \times 1\frac{7}{8} \times \frac{7}{12}$

22 $2\frac{4}{7} \times 4\frac{2}{3} \times 1\frac{1}{4}$

Remember to look for common factors to cancel.

23 $3\frac{2}{3} \times 1\frac{1}{5} \times 1\frac{3}{22}$

24 $1\frac{1}{18} \times 1\frac{4}{5} \times 3\frac{1}{3}$

25 $4\frac{4}{5} \times 1\frac{5}{18} \times 3\frac{3}{4}$

26 $7\frac{1}{2} \times 1\frac{1}{3} \times \frac{9}{10}$

27 $3\frac{1}{5} \times 2\frac{1}{2} \times 1\frac{3}{4}$

28 $4\frac{1}{2} \times 1\frac{1}{7} \times 2\frac{1}{3}$

29 $2\frac{1}{3} \times \frac{6}{11} \times 2\frac{5}{14}$

30 $3\frac{9}{10} \times 1\frac{2}{3} \times 1\frac{3}{13}$

Whole numbers as fractions

A whole number can be written as a fraction with a denominator of 1.

For instance $6 = \frac{6}{1}$. Doing this makes it easier to multiply a whole number by a fraction or a mixed number.

Exercise 6d

Find $6 \times 7\frac{1}{3}$

$$6 \times 7\frac{1}{3} = \frac{6}{1} \times \frac{22}{3}$$
$$= \frac{\overset{2}{6} \times 22}{1 \times \underset{1}{3}} \quad \text{(cancel by 3)}$$
$$= \frac{2 \times 22}{1 \times 1} = 44$$

Find:

1   $5 \times 4\frac{3}{5}$

2   $2\frac{1}{7} \times 14$

3   $3\frac{1}{8} \times 4$

4   $4\frac{1}{6} \times 9$

5   $18 \times 6\frac{1}{9}$

6   $4 \times 3\frac{3}{8}$

7 $3\frac{3}{5} \times 10$ **9** $5\frac{5}{7} \times 21$ **11** $1\frac{3}{4} \times 8$

8 $2\frac{5}{6} \times 3$ **10** $3 \times 6\frac{1}{9}$ **12** $28 \times 1\frac{4}{7}$

Fractions of quantities

Exercise 6e

Find three-fifths of 95 metres.

Remember that 'of' means 'multiply' so to find $\frac{3}{5}$ of 95 you have to

multiply $\frac{3}{5}$ by 95. This is the same as multiplying by $\frac{95}{1}$:

$$\frac{3}{5} \times \frac{95}{1} = \frac{3 \times 95}{5 \times 1}$$

$$= \frac{3 \times 19}{1 \times 1} \text{ (cancelling 5)}$$

$$= 57$$

$\frac{3}{5}$ of 95 metres is 57 metres.

Find three quarters of $1.

You need to convert $1 into cents first.

$1 = 100 cents

$$\frac{3}{4} \times \frac{100}{1} = 75$$

$\frac{3}{4}$ of $1 is 75 cents.

Find:

1 $\frac{1}{3}$ of 18

2 $\frac{1}{5}$ of 30

3 $\frac{1}{7}$ of 21

4 $\frac{2}{3}$ of 24

5 $\frac{5}{7}$ of 14

6 $\frac{1}{4}$ of 24

7 $\frac{1}{6}$ of 30

8 $\frac{1}{8}$ of 64

9 $\frac{5}{6}$ of 36

10 $\frac{3}{8}$ of 40

11 $\frac{3}{5}$ of 20 metres

12 $\frac{5}{9}$ of 45 dollars

13 $\frac{9}{10}$ of 50 litres

14 $\frac{3}{8}$ of 88 miles

15 $\frac{7}{16}$ of 48 gallons

16 $\frac{4}{9}$ of 18 metres

17 $\frac{5}{8}$ of 16 dollars

18 $\frac{4}{9}$ of 63 litres

19 $\frac{3}{7}$ of 35 miles

20 $\frac{8}{11}$ of 121 gallons

21 $\frac{1}{4}$ of $2

22 $\frac{2}{9}$ of 36 cents

23 $\frac{3}{10}$ of $1

24 $\frac{2}{7}$ of 42 cents

25 $\frac{4}{5}$ of 1 year (365 days)

26 $\frac{3}{8}$ of 1 day (24 hours)

27 $\frac{1}{7}$ of 1 week

28 $\frac{1}{3}$ of $9

29 $\frac{3}{5}$ of $1

30 $\frac{7}{8}$ of 1 day (24 hours)

Dividing by fractions

When we divide 6 by 3 we are finding how many threes there are in 6 and we say $6 \div 3 = 2$.

In the same way, when we divide 10 by $\frac{1}{2}$ we are finding how many halves there are in 10; we know that in 1 whole number there are 2 halves.

i.e. $\frac{1}{1} \div \frac{1}{2} = 2$

but $\frac{1}{1} \times \frac{2}{1} = 2$

$\therefore \frac{1}{1} \div \frac{1}{2} = \frac{1}{1} \times \frac{2}{1}$

Also, in 2 wholes there are 4 halves.

i.e. $\frac{2}{1} \div \frac{1}{2} = 4$, also $\frac{2}{1} \times \frac{2}{1} = 4$

$\therefore \frac{2}{1} \div \frac{1}{2} = \frac{2}{1} \times \frac{2}{1}$

Continuing in this way, we see that in 10 wholes there are 20 halves.

i.e. $\frac{10}{1} \div \frac{1}{2} = 20$, and $\frac{10}{1} \times \frac{2}{1} = 20$

$\therefore \frac{10}{1} \div \frac{1}{2} = \frac{1}{1} \quad \frac{2}{1}$

From the above examples, we divided by a fraction by 'turning the fraction upside down' (inverting it) and then multiplying.

This rule holds for division by fractions.

To divide by a fraction we turn that fraction upside down and multiply.

Exercise 6f

1. How many $\frac{1}{2}$ s are there in 7?
2. How many $\frac{1}{4}$ s are there in 5?
3. How many times does $\frac{1}{7}$ go into 3?
4. How many $\frac{3}{5}$ s are there in 9?
5. How many times does $\frac{2}{3}$ go into 8?

Find:

6. $8 \div \frac{4}{5}$
7. $18 \div \frac{6}{7}$
8. $40 \div \frac{8}{9}$
9. $72 \div \frac{8}{11}$
10. $28 \div \frac{14}{15}$
11. $15 \div \frac{5}{6}$

Remember that there are 2 halves in 1 so there are 7×2 halves in 7.

Write 8 as $\frac{8}{1}$, and remember that to divide by a fraction, turn the fraction upside down and multiply. Look for any cancelling before doing the multiplication.

12 $14 \div \frac{7}{8}$

13 $35 \div \frac{5}{7}$

14 $44 \div \frac{4}{9}$

15 $27 \div \frac{9}{13}$

16 $36 \div \frac{4}{7}$

17 $34 \div \frac{17}{19}$

18 $\frac{21}{32} \div \frac{7}{8}$

19 $\frac{9}{25} \div \frac{3}{10}$

20 $\frac{3}{56} \div \frac{9}{14}$

21 $\frac{21}{22} \div \frac{7}{11}$

22 $\frac{8}{75} \div \frac{4}{15}$

23 $\frac{35}{42} \div \frac{5}{6}$

24 $\frac{28}{27} \div \frac{4}{9}$

25 $\frac{22}{45} \div \frac{11}{15}$

26 $\frac{15}{26} \div \frac{5}{13}$

27 $\frac{49}{50} \div \frac{7}{10}$

28 $\frac{8}{21} \div \frac{4}{7}$

29 $\frac{9}{26} \div \frac{12}{13}$

Dividing by whole numbers and mixed numbers

If we want to divide 3 by 5 we can say

$$3 \div 5 = \frac{3}{1} \div \frac{5}{1}$$
$$= \frac{3}{1} \times \frac{1}{5}$$
$$= \frac{3}{5}$$

So $3 \div 5$ is the same as $\frac{3}{5}$. Similarly $7 \div 11$ is $\frac{7}{11}$.

Division with mixed numbers can be done as long as all the mixed numbers are first changed into improper fractions. For example, if we want to divide $3\frac{1}{8}$ by $8\frac{3}{4}$ we first change $3\frac{1}{8}$ into $\frac{25}{8}$ and $8\frac{3}{4}$ into $\frac{35}{4}$. Then we can use the same method as before.

Exercise 6g

Find the value of $3\frac{1}{8} \div 8\frac{3}{4}$.

$3\frac{1}{8} \div 8\frac{3}{4} = \frac{25}{8} \div \frac{35}{4}$ (Express each mixed number as an improper fraction.)

$\qquad = \frac{25^{\,5}}{8^{\,2}} \times \frac{4^{\,1}}{35^{\,7}}$ (To divide by a fraction turn it upside down and multiply.)

$\qquad = \frac{5}{14}$

This flow chart summarises the process.

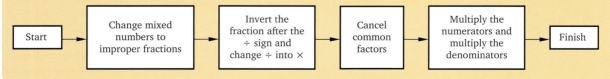

Find:

1 $5\frac{4}{9} \div \frac{14}{27}$

2 $3\frac{1}{8} \div 3\frac{3}{4}$

3 $7\frac{1}{5} \div 1\frac{7}{20}$

4 Divide $8\frac{1}{4}$ by $1\frac{3}{8}$

5 Divide $6\frac{2}{3}$ by $2\frac{4}{9}$

6 $4\frac{2}{7} \div \frac{9}{14}$

7 $5\frac{5}{8} \div 6\frac{1}{4}$

8 $6\frac{4}{9} \div 1\frac{1}{3}$

9 Divide $5\frac{1}{4}$ by $2\frac{11}{12}$

10 Divide $7\frac{1}{7}$ by $1\frac{11}{14}$

11 $10\frac{2}{3} \div 1\frac{7}{9}$

12 $8\frac{4}{5} \div 3\frac{3}{10}$

13 Divide $11\frac{1}{4}$ by $\frac{15}{16}$

14 Divide $9\frac{1}{7}$ by $1\frac{11}{21}$

15 $31\frac{1}{2} \div 5\frac{5}{8}$

16 $9\frac{3}{4} \div 1\frac{5}{8}$

17 $12\frac{1}{2} \div 8\frac{3}{4}$

18 Divide $10\frac{5}{6}$ by $3\frac{1}{4}$

19 Divide $8\frac{2}{3}$ by $5\frac{7}{9}$

20 $22\frac{2}{3} \div 1\frac{8}{9}$

! Investigation

Investigate what happens to a number when you multiply it by a fraction that is less than 1.

Does it matter whether the number itself is more or less than 1?

Mixed multiplication and division

Suppose we want to find the value of an expression like $2\frac{1}{4} \times \frac{3}{14} \div 1\frac{2}{7}$.

Two things need to be done:

Step one – If there are any mixed numbers, change them into improper fractions.

Step two – Turn the fraction *after* the ÷ sign upside down and change ÷ into ×.

Then

$2\frac{1}{4} \times \frac{3}{14} \div 1\frac{2}{7} = \frac{9}{4} \times \frac{3}{14} \div \frac{9}{7}$ Mixed numbers changed to improper fractions

$= \frac{9}{4} \times \frac{3}{14} \times \frac{7}{9}$ $\frac{9}{7}$ turned upside down and ÷ changed to ×

$= \frac{3}{8}$

Exercise 6h

Find:

1 $\frac{5}{8} \times 1\frac{1}{2} \div \frac{15}{16}$

2 $2\frac{3}{4} \times \frac{5}{6} \div \frac{11}{12}$

3 $\frac{2}{3} \times 1\frac{1}{5} \div \frac{12}{25}$

4 $\frac{4}{7} \times \frac{8}{9} \div \frac{16}{21}$

5 $\frac{2}{5} \times \frac{9}{10} \div \frac{27}{40}$

6 $\frac{3}{4} \times 2\frac{1}{3} \div \frac{21}{32}$

7 $3\frac{2}{5} \times \frac{4}{5} \div \frac{8}{15}$

8 $\frac{3}{7} \times 2\frac{1}{2} \div \frac{10}{21}$

9 $\frac{3}{5} \times \frac{9}{11} \div \frac{18}{55}$

10 $\frac{1}{4} \times \frac{11}{12} \div \frac{22}{27}$

11 $\frac{3}{7} \times \frac{2}{5} \div \frac{8}{21}$

12 $\frac{14}{25} \times \frac{5}{9} \div \frac{7}{18}$

Mixed operations

When brackets are placed round a pair of fractions it means that we have to work out what is *inside* the brackets before doing anything else. For example

$$\left(\tfrac{1}{2}+\tfrac{1}{4}\right)\times\tfrac{5}{7}=\left(\tfrac{2+1}{4}\right)\times\tfrac{5}{7}$$

$$=\tfrac{3}{4}\times\tfrac{5}{7}$$

$$=\tfrac{15}{28}$$

If we meet an expression in which $+$, $-$, $\times$ and $\div$ occur, we need to know the order in which to do the calculations. We use the same rule for fractions as we used for whole numbers, that is

Brackets first, then Multiply and Divide, then Add and Subtract.

You may remember this order from the phrase

'Bless My Dear Aunt Sally'.

Exercise 6i

Find $\tfrac{2}{5}-\tfrac{1}{2}\times\tfrac{3}{5}+\tfrac{1}{10}$

$$\tfrac{2}{5}-\tfrac{1}{2}\times\tfrac{3}{5}+\tfrac{1}{10}=\tfrac{2}{5}-\tfrac{3}{10}+\tfrac{1}{10} \qquad \text{(multiply)}$$

$$=\tfrac{4-3+1}{10} \qquad \text{(add and subtract)}$$

$$=\tfrac{2^{1}}{10_{5}}$$

$$=\tfrac{1}{5}$$

Calculate:

1 $\tfrac{1}{2}+\tfrac{1}{4}\times\tfrac{2}{5}$

2 $\tfrac{2}{3}\times\tfrac{1}{2}+\tfrac{1}{4}$

3 $\tfrac{4}{5}-\tfrac{3}{10}\div\tfrac{1}{2}$

4 $\tfrac{2}{7}\div\tfrac{2}{3}-\tfrac{3}{14}$

5 $\tfrac{4}{5}+\tfrac{3}{10}\times\tfrac{2}{9}$

6 $\tfrac{1}{3}-\tfrac{1}{2}\times\tfrac{1}{4}$

7 $\tfrac{3}{4}\div\tfrac{1}{2}+\tfrac{1}{8}$

8 $\tfrac{1}{7}+\tfrac{5}{8}\div\tfrac{3}{4}$

9 $\tfrac{5}{6}\times\tfrac{3}{10}-\tfrac{3}{16}$

10 $\tfrac{3}{7}-\tfrac{1}{4}\times\tfrac{8}{21}$

11 $\left(\tfrac{4}{9}-\tfrac{1}{3}\right)\times\tfrac{6}{7}$

12 $\tfrac{3}{5}\times\left(\tfrac{2}{3}+\tfrac{1}{2}\right)$

13 $\tfrac{7}{8}\div\left(\tfrac{3}{4}+\tfrac{2}{3}\right)$

14 $\left(\tfrac{3}{10}+\tfrac{2}{5}\right)\div\tfrac{7}{15}$

15 $\left(\tfrac{5}{11}-\tfrac{1}{3}\right)\times\tfrac{3}{8}$

Work out the value inside the brackets first.

16 $\frac{3}{8} \div \left(\frac{2}{3} + \frac{1}{4}\right)$

17 $\left(\frac{4}{7} + \frac{1}{3}\right) \div 3\frac{4}{5}$

18 $\frac{5}{9} \times \left(\frac{2}{3} - \frac{1}{6}\right)$

19 $\left(\frac{6}{11} - \frac{1}{2}\right) \div \frac{3}{4}$

20 $\frac{9}{10} \div \left(\frac{1}{6} + \frac{2}{3}\right)$

21 $\frac{1}{6} \times \left(\frac{2}{3} - \frac{1}{2}\right) \div \frac{7}{12}$

22 $\frac{7}{10} \div \left(\frac{2}{5} + \frac{4}{15} \times \frac{3}{5}\right)$

23 $\left(2\frac{1}{4} + \frac{3}{8}\right) \times \frac{2}{3} - 1\frac{1}{2}$

24 $1\frac{3}{11} - \frac{6}{7} \times 1\frac{5}{9} + \frac{13}{33}$

25 $\frac{5}{8} \times \left(\frac{4}{9} - \frac{1}{6}\right) \div 1\frac{9}{16}$

26 $\frac{2}{9} + \left(\frac{6}{7} \div \frac{3}{4}\right) \times 3\frac{1}{2}$

27 $1\frac{1}{10} \times \frac{23}{24} \div \left(\frac{3}{5} + \frac{1}{6}\right)$

28 $2\frac{2}{5} - \frac{7}{10} \times \left(\frac{4}{7} - \frac{1}{3}\right)$

29 $\frac{5}{9} \div \left(1\frac{1}{3} + \frac{4}{9}\right) + \frac{3}{8}$

30 $1\frac{1}{9} + \left(\frac{5}{6} - \frac{3}{4} \div 4\frac{1}{2}\right)$

State whether each of the following statements is true or false:

31 $\frac{1}{2} \times \frac{2}{3} + \frac{1}{3} = \frac{1}{3} + \frac{1}{3}$

32 $\frac{1}{3} \times \frac{3}{4} + \frac{1}{4} = \frac{1}{3} \times 1$

33 $\frac{1}{4} \div \frac{3}{4} + \frac{1}{2} = \frac{1}{3} + \frac{1}{2}$

34 $\frac{1}{3} + \frac{2}{3} \times \frac{1}{4} = \frac{1}{3} + \frac{1}{6}$

35 $\frac{1}{2} + \frac{1}{4} \div \frac{1}{2} = \frac{3}{4} \times \frac{2}{1}$

36 $\frac{3}{4} - \frac{1}{2} \times \frac{2}{3} = \frac{1}{4} \times \frac{2}{3}$

37 $\frac{2}{3} - \frac{1}{4} + \frac{1}{2} = \frac{2}{3} + \frac{1}{2} - \frac{1}{4}$

38 $\frac{3}{5} \times \frac{2}{3} + \frac{1}{2} = \frac{3}{5} + \frac{1}{2} \times \frac{2}{3}$

39 $\frac{4}{7} - \frac{1}{4} \div \frac{1}{3} = \left(\frac{4}{7} - \frac{1}{4}\right) \div \frac{1}{3}$

40 $\frac{3}{8} \div \frac{1}{4} - \frac{1}{4} = \frac{3}{8} \times \frac{4}{1} - \frac{1}{4}$

Exercise 6j

In this exercise you will find +, −, × and ÷. Read the question carefully and then decide which method to use. Find:

1 $1\frac{1}{2} + 3\frac{1}{4}$

2 $2\frac{3}{8} - 1\frac{1}{4}$

3 $1\frac{1}{5} \times \frac{5}{8}$

4 $3\frac{1}{2} \div \frac{7}{8}$

5 $\frac{4}{7} + 1\frac{1}{2}$

6 $4\frac{1}{4} \times \frac{2}{9}$

7 $3\frac{2}{3} \div \frac{1}{6}$

8 $2\frac{1}{5} - 1\frac{1}{3}$

9 $5\frac{1}{2} \times \frac{6}{11}$

10 $1\frac{3}{8} + 2\frac{1}{2}$

11 $5\frac{1}{2} + \frac{3}{4}$

12 $4\frac{1}{3} \times \frac{6}{13}$

13 $3\frac{4}{5} - 2\frac{1}{10}$

14 $3\frac{1}{7} \div 1\frac{3}{8}$

15 $4\frac{1}{5} \times \frac{4}{7}$

16 $2\frac{5}{6} \div 3\frac{1}{3}$

17 $1\frac{4}{7} + 2\frac{1}{2}$

18 $2\frac{3}{4} - 1\frac{7}{8}$

19 $2\frac{3}{8} + 1\frac{7}{16}$

20 $5\frac{1}{4} \div 1\frac{1}{6}$

21 $1\frac{1}{4} + \frac{2}{3} - \frac{5}{6}$

22 $2\frac{1}{2} - \frac{2}{3} - 1\frac{1}{4}$

23 $3\frac{1}{2} + 1\frac{1}{4} - \frac{5}{8}$

24 $2\frac{1}{3} + 1\frac{1}{2} - \frac{3}{4}$

25 $4\frac{1}{8} - 5\frac{3}{4} + 2\frac{1}{2}$

26 $4\frac{1}{2} - 5\frac{1}{4} + 2\frac{1}{8}$

27 $3\frac{4}{5} + \frac{3}{10} - 1\frac{1}{20}$

28 $5\frac{1}{2} - 1\frac{3}{4} - 2\frac{1}{4}$

29 $3\frac{1}{7} + 2\frac{1}{2} - \frac{3}{14}$

30 $5\frac{1}{2} - \frac{3}{4} - 4\frac{1}{4}$

31 $1\frac{1}{2} + 2\frac{2}{3} \times \frac{3}{4}$

32 $2\frac{1}{3} \times 1\frac{1}{2} - 2\frac{1}{2}$

33 $4\frac{1}{3} \div 2\frac{1}{6} + \frac{1}{4}$

34 $2\frac{2}{5} - \frac{6}{7} \div \frac{5}{14}$

35 $1\frac{1}{8} \times \frac{4}{9} \div 2\frac{1}{2}$

36 $2\frac{3}{8} - 1\frac{1}{5} \times 1\frac{2}{3}$

37 $1\frac{3}{4} \div 4\frac{2}{3} - \frac{5}{16}$

38 $2\frac{1}{7} \times 3\frac{1}{4} \div 1\frac{5}{8}$

39 $1\frac{1}{2} + 2\frac{5}{7} - 1\frac{5}{14}$

40 $\frac{3}{5} \times 1\frac{1}{4} \div \frac{3}{8}$

Problems

Exercise 6k

If Jane can iron a shirt in $4\frac{3}{4}$ minutes, how long will it take her to iron 10 shirts?

You know the time to iron 1 shirt. It will take 10 times as long to iron 10 shirts.

Time to iron $\quad$ 1 shirt $= 4\frac{3}{4}$ minutes

Time to iron 10 shirts $= 4\frac{3}{4} \times 10$ minutes

$$= \frac{19}{4} \times \frac{10}{1} \text{ minutes}$$

$$= \frac{95}{2} \text{ minutes}$$

$$= 47\frac{1}{2} \text{ minutes}$$

A piece of string of length $22\frac{1}{2}$ cm is to be cut into small pieces each $\frac{3}{4}$ cm long. How many pieces can be obtained?

Number of small pieces = length of string ÷ length of one short piece

You want to find the number of $\frac{3}{4}$ cm pieces in $22\frac{1}{2}$ cm so you need to find $22\frac{1}{2} \div \frac{3}{4}$.

Number of small pieces $= 22\frac{1}{2} \div \frac{3}{4}$

(Express $22\frac{1}{2}$ as an improper fraction and multiply by $\frac{3}{4}$ turned upside down.)

$$= \frac{\overset{15}{\cancel{45}} \times \overset{2}{\cancel{4}}}{\underset{1}{\cancel{2}} \times \underset{1}{\cancel{3}}} = 30$$

Thus 30 pieces can be obtained.

1 A bag of flour weighs $1\frac{1}{2}$ kilograms. What is the weight of 20 bags?

2 A cook adds $3\frac{1}{2}$ cups of water to a stew. If the cup holds $\frac{1}{10}$ of a litre how many litres of water were added?

3 My journey to school starts with a walk of $\frac{1}{2}$ km to the bus stop, then a bus ride of $2\frac{1}{5}$ km followed by a walk of $\frac{3}{10}$ km. How long is my journey to school?

4 It takes $3\frac{1}{4}$ minutes for a cub scout to clean a pair of shoes. If he cleans 18 pairs of shoes to raise money for charity, how long does he spend on the job?

5   A burger bar chef cooks some beefburgers and piles them one on top of the other. If each burger is $9\frac{1}{2}$ mm thick and the pile is 209 mm high, how many did he cook?

6   If you read 30 pages of a book in $\frac{3}{4}$ of an hour, how many minutes does it take to read each page?

Mixed exercises

Exercise 6I

1   Calculate
     a $\frac{3}{4}+\frac{11}{12}$
     b $3\frac{1}{8}-2\frac{1}{4}+1\frac{1}{2}$

2   Find how many times $2\frac{1}{4}$ goes into $13\frac{1}{2}$.

3   What is $\frac{7}{9}$ of $1\frac{1}{14}$?

4   Find $\frac{3}{5}+1\frac{1}{2}\times\frac{7}{10}$

5   Arrange the following fractions in ascending order of size: $\frac{7}{10}, \frac{3}{5}, \frac{2}{3}$

6   Find:
     a $4\frac{1}{7}\times4\frac{2}{3}$
     b $3\frac{3}{8}\div2\frac{1}{4}$

7   What is $\frac{3}{4}$ of $\frac{8}{9}$ added to $1\frac{1}{2}$?

8   Find $\left(1\frac{7}{8}+2\frac{1}{4}\right)\times1\frac{5}{11}$

9   What is $\frac{2}{7}$ of 1 hour 3 minutes (in minutes)?

10   Find $7\frac{1}{5}-4\frac{1}{8}\div1\frac{1}{4}$

11   Fill in the missing numbers:
     a $\frac{7}{9}=\frac{21}{}$
     b $\frac{10}{11}=\frac{}{44}$

12   Express as mixed numbers:
     a $\frac{13}{5}$
     b $\frac{31}{8}$
     c $\frac{27}{5}$

13   State whether the following statements are true or false:
     a $\frac{4}{11}>\frac{3}{10}$
     b $\frac{3}{7}$ of $5=\frac{3}{7}\times\frac{5}{1}$
     c $2\frac{1}{7}=\frac{7}{15}$

14   A handyman takes $1\frac{1}{8}$ minutes to lay one brick. How long will it take him to lay 56 bricks?

15   A pharmacist counts 48 tablets and puts them in a bottle. Each tablet weighs $\frac{1}{4}$ of a gram and the weight of the empty bottle is $112\frac{1}{2}$ grams. What is the total weight?

1   Find:
    a $4\frac{1}{2} \times 3\frac{1}{3}$ **b** $3\frac{2}{5} \div \frac{3}{10}$

2   Find:
    a $\frac{8}{9} + \frac{21}{27}$ **b** $2\frac{1}{3} + \frac{4}{9} + 1\frac{5}{6}$

3   Put > or < between the following pairs of numbers:
    a $\frac{4}{7}$ $\frac{5}{8}$ **b** $\frac{11}{9}$ $1\frac{3}{10}$

4   Calculate:
    a $5\frac{1}{4} - 1\frac{2}{3} \div \frac{2}{5}$ **b** $3\frac{3}{8} \times \left(8\frac{1}{2} - 5\frac{5}{6}\right)$

5   Arrange in ascending order: $\frac{7}{15}, \frac{1}{3}, \frac{2}{5}$

6   What is $1\frac{1}{2}$ subtracted from $\frac{2}{3}$ of $5\frac{1}{4}$?

7   Find:
    a $4\frac{1}{2} \times 3\frac{2}{3} - 10\frac{1}{4}$ **b** $3\frac{1}{2} \div \left(2\frac{1}{8} - \frac{3}{4}\right)$

8   What is $1\frac{2}{3}$ of 1 minute 15 seconds (in seconds)?

9   Fill in the missing numbers:
    a $\frac{4}{5} = \frac{}{30}$ **b** $\frac{2}{7} = \frac{6}{}$

10 Express as mixed numbers:
    a $\frac{25}{8}$ **b** $\frac{49}{9}$ **c** $\frac{37}{6}$

11 A car travels $5\frac{1}{4}$ km north, then $2\frac{1}{2}$ km west and finally $4\frac{3}{8}$ km north. What is the total distance travelled (in kilometres)? What fraction of the journey was travelled in a northerly direction?

12 A man can paint a door in 1 hour 15 minutes. How many similar doors can he paint in $7\frac{1}{2}$ hours?

1   Find:
    a $1\frac{5}{6} + \frac{5}{18} + \frac{7}{12}$ **b** $1\frac{2}{3} - 2\frac{1}{5} + \frac{8}{15}$

2   Find:
    a $1\frac{5}{6} \div 7\frac{1}{3}$ **b** $2\frac{1}{4} \times \frac{16}{45}$

3   What is $\frac{5}{6}$ of the number of days in June?

4   Arrange in descending order: $\frac{17}{20}, \frac{3}{4}, \frac{7}{10}$

5 Calculate:

   a $4\frac{1}{2} \times 3\frac{2}{3} - 10\frac{1}{4}$ **b** $3\frac{1}{4} + 5\frac{1}{2} \div \frac{3}{8}$

6 Which is smaller, $\frac{8}{11}$ or $\frac{7}{9}$?

7 Find $3\frac{9}{10} \div \left(3\frac{3}{5} - 1\frac{1}{2}\right)$

8 What is $\frac{4}{7}$ of $4\frac{2}{3}$ divided by $1\frac{1}{9}$?

9 Express as mixed numbers:

   a $\frac{22}{3}$ **b** $\frac{46}{5}$ **c** $\frac{106}{10}$

10 Which of the following statements are true?

   a $3\frac{1}{2} \div 1 = 3\frac{1}{2} \times 1$ **b** $\frac{1}{2} \times \left(\frac{1}{4} + \frac{1}{8}\right) =$ half of $\frac{3}{8}$ **c** $\frac{1}{2} + \frac{1}{2} \div 2 = \frac{3}{4}$

11 It takes $1\frac{3}{4}$ minutes to wrap a parcel and a half a minute to address it.
How long does it take to wrap and address 8 similar parcels?

12 My bag contains 2 books each of weight $\frac{3}{7}$ kg and 3 folders each
of weight $\frac{5}{21}$ kg.
What is the total weight in my bag? What fraction of the total weight
is books?

(?) Puzzle

Gary spent one-third of his money at the Sports Club and two-thirds of
the remainder at the supermarket. He had $12 left. How much did he
have to start with?

In this chapter you have seen that...

✔ 'of' means 'multiplied by'

✔ you can multiply and divide whole numbers by fractions by writing
the whole number over 1

✔ to multiply fractions by fractions you multiply the numerators
together and you multiply the denominators together

✔ to divide by a fraction you turn the fraction upside down and multiply

✔ you can multiply and divide with mixed numbers by turning the mixed
numbers into improper fractions.

7 Decimals

At the end of this chapter you should be able to...

1. Write a given number in expanded form – under the heading hundreds, tens, units, etc.
2. Write a decimal number as a fraction and vice versa.
3. Add and subtract decimal numbers.
4. Multiply and divide decimal numbers by 10, 100, 1000, ...
5. Multiply and divide decimal numbers by whole numbers.
6. Multiply two decimal numbers.
7. Solve problems using operations on decimal numbers.

You need to know...

✔ how to add and subtract whole numbers
✔ how to multiply whole numbers
✔ how to multiply fractions together
✔ how to do short and long division.

Key words

decimal, fraction, perimeter, triangle, product, quadrilateral, rectangle, regular pentagon.

The meaning of decimals

Consider the number 426. The position of the digits indicates what each digit represents. We can write:

hundreds	tens	units
4	2	6

Each quantity in the heading is $\frac{1}{10}$ of the quantity to its left: ten is $\frac{1}{10}$ of a hundred, a unit is $\frac{1}{10}$ of ten. Moving further to the right we can have further headings: tenths of a unit, hundredths of a unit and so on (a hundredth of a unit is $\frac{1}{10}$ of a tenth of a unit).

For example

tens	units		tenths	hundredths
1	6	.	0	2

To mark where the units come we put a point after the units position.

16.02 is 1 ten, 6 units and 2 hundredths or $16\frac{2}{100}$.

units		tenths	hundredths	thousandths
0	.	0	0	4

0.004 is 4 thousandths or $\frac{4}{1000}$. In this case, 0 is written before the point to help make it clear where the point comes.

Ordering decimals

You can compare the sizes of decimals by looking at the digits in the place values, starting with the highest place value.

For example, 6.277 is smaller than 6.3 because the units are the same, but 2 tenths is smaller than 3 tenths.

Exercise 7a

Write the following numbers in headed columns:

	tens	units		tenths	hundredths
34.62 =	3	4	.	6	2

	units		tenths	hundredths	thousandths	ten-thousandths
0.0207 =	0	.	0	2	0	7

1	2.6	**4**	0.09	**7**	1.046	**10**	0.604
2	32.1	**5**	101.3	**8**	12.001	**11**	15.045
3	6.03	**6**	0.00007	**9**	6.34	**12**	0.0092

Write these numbers in order of size with the largest number first.

13	5.86, 4.99	**15**	38.64, 38.46
14	2.27, 2.29	**16**	6, 5.98

Write these numbers in order of size with the smallest number first.

17	9.45, 8.99	**19**	15.35, 15.53
18	3.74, 3.71	**20**	6.02, 6

Write these numbers in order of size with the smallest number first.

21 6.72, 2.81, 6.27   **23** 0.55, 5.5, 0.505, 0.055

22 7.07, 7.41, 7.18, 7.03

Write these numbers in order of size with the largest number first.

24 10.02, 14.16, 12.32, 13.55   **26** 0.11, 0.011, 0.101, 0.111

25 6.555, 6.5, 6.05, 6.55

Changing decimals to fractions

Exercise 7b

Write the following decimals as fractions in their lowest terms (using mixed numbers where necessary):

	units		tenths		
$0.6 =$	0	.	6	$= \dfrac{6}{10}$	Now cancel.
				$= \dfrac{3}{5}$	

	tens	units		tenths	hundredths		
$12.04 =$	1	2	.	0	4	$= 12\dfrac{4}{100}$	Now cancel the fraction.
						$= 12\dfrac{1}{25}$	

1	0.2	**4**	0.0007	**7**	0.7	**10**	1.7
2	0.06	**5**	0.001	**8**	2.01	**11**	15.5
3	1.3	**6**	6.4	**9**	1.8	**12**	8.06

You can go straight from the decimal to one fraction.

0.302

	units	tenths	hundredths	thousandths	
$0.302 =$	0 .	3	0	2	$= \dfrac{3}{10} + \dfrac{2}{1000}$

Now write these with a common denominator. $= \dfrac{300}{1000} + \dfrac{2}{1000}$

     $= \dfrac{302}{1000}$

Now cancel. $= \dfrac{151}{500}$

You can miss out the first two steps and go straight to one fraction.

Write as fractions:

13	0.73	**16**	0.0029	**19**	0.071	**22**	0.63
14	0.081	**17**	0.00067	**20**	0.3001	**23**	0.031
15	0.207	**18**	0.17	**21**	0.0207	**24**	0.47

Write as fractions in their lowest terms:

25	0.25	**28**	0.0305	**31**	0.35	**34**	0.125
26	0.072	**29**	0.15	**32**	0.0016	**35**	0.48
27	0.38	**30**	0.025	**33**	0.044	**36**	0.625

Changing fractions to decimals

Exercise 7c

Write the following numbers as decimals:

$3\frac{3}{100}$

	units	tenths	hundredths
$3\frac{3}{100} =$	3 .	0	3

1 $\frac{7}{100}$ **4** $\frac{2}{1000}$ **7** $\frac{4}{100}$ **10** $\frac{6}{10\,000}$

2 $\frac{9}{10}$ **5** $\frac{4}{10}$ **8** $7\frac{8}{10}$ **11** $4\frac{5}{1000}$

3 $1\frac{1}{10}$ **6** $2\frac{6}{100}$ **9** $7\frac{8}{100}$ **12** $\frac{29}{10\,000}$

Addition of decimals

To add decimals you can write them in columns. It is important to keep the decimal points in line.

	tens	units	tenths	
$4.2 + 13.1 = 17.3$		4 .	2	2 tenths + 1 tenth = 3 tenths
$+$	1	3 .	1	
	1	7 .	3	

		units	tenths	
$5.3 + 6.8 = 12.1$		5 .	3	3 tenths + 8 tenths = 11 tenths
$+$		6 .	8	= 1 unit and 1 tenth
	1	2 .	1	

The headings above the digits need not be written as long as we know what they are and the decimal points are in line (including the invisible point after a whole number, e.g. 4 = 4.0).

Exercise 7d

Find $3 + 1.6 + 0.032 + 2.0066$

Write the numbers in a column, keeping the decimal points in line.

$3 + 1.6 + 0.032 + 2.0066 = 6.6386$

```
   3
   1.6
   0.032
 +2.0066
  6.6386
```

Find:

1	$7.2 + 3.6$	**11**	$0.0043 + 0.263$	**21**	Add 0.68 to 1.7.
2	$6.21 + 1.34$	**12**	$0.002 + 2.1$	**22**	Find the sum of 3.28 and 14.021.
3	$0.013 + 0.026$	**13**	$0.00052 + 0.00124$		
4	$3.87 + 0.11$	**14**	$0.068 + 0.003 + 0.06$	**23**	To 7.9 add 4 and 3.72.
5	$4.6 + 1.23$	**15**	$4.62 + 0.078$	**24**	Evaluate $7.9 + 0.62 + 5$.
6	$13.14 + 0.9$	**16**	$0.32 + 0.032 + 0.0032$	**25**	Find the sum of 8.6, 5 and 3.21.
7	$4 + 3.6$	**17**	$4.6 + 0.0005$		
8	$9.24 + 3$	**18**	$16.8 + 3.9$		
9	$3.6 + 0.08$	**19**	$1.62 + 2.078 + 3.1$		
10	$7.2 + 0.32 + 1.6$	**20**	$7.34 + 6 + 14.034$		

Remember that 4 is the same as 4.0.

? Puzzle

4.33	0.59	2.36	5.608	3.182	0.57	0.649
6.25	1.89	5.81	3.218	1.14	2.98	3.902
3.72	0.9	3.7	5.989	6.27	6.804	0.098
0.13	5.91	3.241	0.68	1.291	2.99	4.2

1 You have a time limit of two minutes for this exercise.
Pair off as many numbers as possible so that all your number pairs add up to a number between 5 and 7. When your time is up your score is the sum of all the remaining numbers. The lower the score the better. (You will probably find it helpful to copy the list and cross out each pair of numbers that satisfies the condition.)

2 Repeat the exercise to try to reduce your score.

3 What is the lowest score possible?

4 Try to write down twelve numbers written as three rows with four numbers in each row so that the rules given in part **a** apply and the lowest possible score is 0. If you think you've succeeded try it on a friend.

5 Using the same list of decimals pair them off so that the difference between the pairs lies between 0 and 1.

Subtraction of decimals

Exercise 7e

Subtraction may also be done by writing the numbers in columns, making sure that the decimal points are in line.

Find
24.2 – 13.7

$$\begin{array}{r} 24.2 \\ -13.7 \\ \hline 10.5 \end{array}$$

24.2 – 13.7 = 10.5

Find:

1	6.8 – 4.3	**5**	0.0342 – 0.0021	**9**	102.6 – 31.2
2	9.6 – 1.8	**6**	17.23 – 0.36	**10**	7.32 – 0.67
3	32.7 – 14.2	**7**	3.273 – 1.032	**11**	54.07 – 12.62
4	0.62 – 0.21	**8**	0.262 – 0.071	**12**	7.063 – 0.124

It may be necessary to add zeros so that there are the same number of digits after the point in both cases.

4.623 – 1.7   Fill 'empty' places with zeros.

$$\begin{array}{r} 4.623 \\ -1.700 \\ \hline 2.923 \end{array}$$

4.623 – 1.7 = 2.923

4.63 – 1.0342   Fill 'empty' places with zeros.

$$\begin{array}{r} 4.6300 \\ -1.0342 \\ \hline 3.5958 \end{array}$$

4.63 – 1.0342 = 3.5958

13	$3.26 - 0.2$	**22**	$0.000\,32 - 0.000\,123$	**31**	$0.73 - 0.000\,06$
14	$3.2 - 0.26$	**23**	$0.0073 - 0.0006$	**32**	$0.73 - 0.6$
15	$14.23 - 11.1$	**24**	$0.0073 - 0.006$	**33**	Take 19.2 from 76.8.
16	$6.8 - 4.14$	**25**	$0.006 - 0.00073$	**34**	Subtract 1.9 from 10.2.
17	$11 - 8.6$	**26**	$0.06 - 0.00073$	**35**	From 0.168 subtract 0.019.
18	$7.98 - 0.098$	**27**	$6 - 0.73$	**36**	Evaluate $7.62 - 0.81$.
19	$7.098 - 0.98$	**28**	$6 - 0.073$		
20	$3.2 - 0.428$	**29**	$7.3 - 0.06$		
21	$11.2 - 0.0026$	**30**	$730 - 0.6$		

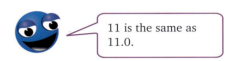

11 is the same as 11.0.

Exercise 7f

Find the value of:

1	$8.62 + 1.7$	**11**	$38.2 + 1.68$
2	$8.62 - 1.7$	**12**	$38.2 - 1.68$
3	$3.8 - 0.82$	**13**	$0.84 + 2 + 200$
4	$0.08 + 0.32 + 6.2$	**14**	$16 + 1.6 + 0.16$
5	$5 - 0.6$	**15**	$1.4 - 0.81$
6	$100 + 0.28$	**16**	$0.02 - 0.013$
7	$100 - 0.28$	**17**	$0.062 + 0.32$
8	$0.26 + 0.026$	**18**	$6.83 - 0.19$
9	$0.26 - 0.026$	**19**	$17.2 + 20 + 1.62$
10	$78.42 - 0.8$	**20**	$9.2 + 13.21 - 14.6$

The perimeter (the distance all round) of the triangle is 6.5 cm.

What is the length of the third side?

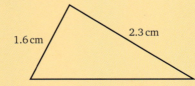

1.6 cm

2.3 cm

As the distance round the three sides is 6.5 cm, you can find the third side by adding the lengths of the two sides that you know, then take the result from 6.5 cm.

1.6 cm + 2.3 cm = 3.9 cm

$$\begin{array}{r} 1.6 \\ +2.3 \\ \hline 3.9 \end{array}$$

The length of the third side is 6.5 cm − 3.9 cm = 2.6 cm

$$\begin{array}{r} 6.5 \\ -3.9 \\ \hline 2.6 \end{array}$$

21 Find the perimeter of the rectangle:

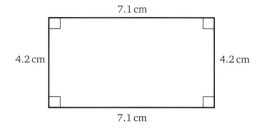

22 A piece of webbing is 7.6 m long. If 2.3 m is cut off, how much is left?

23 Find the total bill for three articles bought in the USA costing US $5, US $6.52 and US $13.25.

24 The bill for two books came to $2848. One book cost $744. What was the cost of the other one?

25 Add 2.32 and 0.68 and subtract the result from 4.

26 Find the perimeter of the quadrilateral:

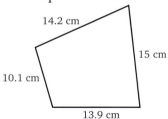

27 The bill for three meals was $3000. The first meal cost $715 and the second $1360. What was the cost of the third?

28 The perimeter of the quadrilateral is 19 cm. What is the length of the fourth side?

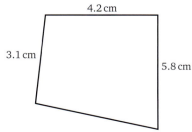

Puzzle

If the sum of the numbers in all the rows, columns and diagonals of a square is the same the square is called a magic square.

9	8	4
2	7	12
10	6	5

For example, in this magic square the total in every row, column and diagonal is 21.

1

		8.1
5.4	6.3	7.2
	10.8	

Fill in the blanks in this magic square if the total is always 18.9.

2

6.3	5.5	2.8
1.4	4.9	8.5
7.0	4.2	3.5

This magic square contains two wrong numbers. Find these wrong numbers and correct them.

3

15	10	30	60
40	50	16	90
14	11	20	70
10	80	13	12

Insert decimal points so that this is a magic square.

Multiplication by 10, 100, 1000, ...

Consider $32 \times 10 = 320$. Writing 32 and 320 in headed columns gives

hundreds	tens	units
	3	2
3	2	0

Multiplying by 10 has made the number of units become the number of tens, and the number of tens has become the number of hundreds, so that all the digits have moved one place to the left.

Consider 0.2×10. When multiplied by 10, tenths become units $\left(\frac{1}{10} \times 10 = 1\right)$, so

units	tenths		units
0	. 2	$\times$ 10	= 2

Again the digit has moved one place to the left.

Multiplying by 100 means multiplying by 10 and then by 10 again, so the digits move 2 places to the left.

tens	units		tenths	hundredths	thousandths	
	0	.	4	2	6	× 100
= 4	2	.	6			

Notice that the digits move to the left while the point stays put but without headings it looks as though the digits stay put and the point moves to the right.

When necessary we fill in an empty space with a nought.

units		tenths			hundreds	tens	units
4	.	2	× 100	=	4	2	0

Exercise 7g

Find the value of:

> **a** 368×100 $368 \times 100 = 36\,800$
>
> **b** 3.68×10 $3.68 \times 10 = 36.8$
>
> **c** 3.68×1000 $3.68 \times 1000 = 3680$

1	72×1000	**5**	32.78×100	**9**	72.81×1000
2	8.24×10	**6**	$0.043 \times 10\,000$	**10**	$0.000\,0063 \times 10$
3	0.0024×100	**7**	0.0602×1000	**11**	$0.007\,03 \times 100$
4	46×10	**8**	3.206×10	**12**	$0.0374 \times 10\,000$

Division by 10, 100, 1000, ...

When we divide by 10, hundreds become tens and tens become units.

hundreds	tens	units			tens	units
6	4	0	÷ 10	=	6	4

The digits move one place to the right and the number becomes smaller but it looks as though the decimal point moves to the left so

$$2.72 \div 10 = 0.272$$

To divide by 100 the point is moved two places to the left.

To divide by 1000 the point is moved three places to the left.

Find the value of:

> **a** $3.2 \div 10$ $3.2 \div 10 = 0.32$
>
> **b** $320 \div 10\,000$
>
> The units become tenthousandths, the tens become thousandths, and so on.
>
> $320 \div 10\,000 = 0.0320 = 0.032$
>
> The final zero can be omitted because it doesn't affect the value of anything.

1	$277.2 \div 100$	**5**	$27 \div 10$	**9**	$426 \div 10\,000$
2	$76.26 \div 10$	**6**	$6.8 \div 100$	**10**	$13.4 \div 10$
3	$0.000\,24 \div 10$	**7**	$0.26 \div 10$	**11**	$3.74 \div 1000$
4	$1.4 \div 100$	**8**	$15.8 \div 1000$	**12**	$0.92 \div 100$

Mixed multiplication and division

Find:

1	$1.6 \div 10$	**9**	$140 \div 1000$	**17**	11.1×1000
2	1.6×10	**10**	$7.8 \times 10\,000$	**18**	$0.038 \div 100$
3	0.078×100	**11**	$24 \div 100$	**19**	$0.38 \div 100$
4	$0.078 \div 100$	**12**	0.063×1000	**20**	$3.8 \times 100\,000$
5	14.2×100	**13**	0.32×10	**21**	$0.024 \div 100$
6	0.068×100	**14**	$7.9 \div 100$	**22**	$0.3 \div 100\,000$
7	$1.63 \div 100$	**15**	0.00078×100	**23**	0.0041×1000
8	$2 \div 1000$	**16**	$2.4 \div 10$	**24**	0.1004×100

25 Share 42 m of string equally amongst 10 people.

26 Find the total cost of 100 articles at $63.52 each.

27 Evaluate $13.8 \div 100$ and 13.8×100.

28 Multiply 1.6 by 100 and then divide the result by 1000.

29 Add 16.2 and 1.26 and divide the result by 100.

30 Take 9.6 from 13.4 and divide the result by 1000.

Division by whole numbers

We can see that

units		tenths			units		tenths
0	.	6	$\div 2$	=	0	.	3

because 6 tenths $\div 2 = 3$ tenths. So we may divide by a whole number using the same layout as we do with whole numbers as long as we keep the digits in the correct columns and the points are in line.

Exercise 7j

Find the value of:

1	$0.4 \div 2$	**4**	$7.8 \div 3$	**7**	$0.672 \div 3$	**10**	$7.53 \div 3$
2	$3.2 \div 2$	**5**	$0.9 \div 9$	**8**	$26.6 \div 7$	**11**	$6.56 \div 4$
3	$0.63 \div 3$	**6**	$0.95 \div 5$	**9**	$42.6 \div 2$	**12**	$0.75 \div 5$

It may sometimes be necessary to fill spaces with noughts.

$0.00036 \div 3$

$0.00036 \div 3 = 0.00012$

$$3\overline{)0.00036}$$
$$0.00012$$

$0.45 \div 5$

$0.45 \div 5 = 0.09$

$$5\overline{)0.45}$$
$$0.09$$

$6.12 \div 3$

$6.12 \div 3 = 2.04$

$$3\overline{)6.12}$$
$$2.04$$

13	$0.057 \div 3$	**17**	$0.012 \div 6$	**21**	$1.232 \div 4$
14	$0.00065 \div 5$	**18**	$0.00036 \div 6$	**22**	$0.6552 \div 6$
15	$0.00872 \div 4$	**19**	$1.62 \div 2$	**23**	$0.0285 \div 5$
16	$0.168 \div 4$	**20**	$4.24 \div 4$	**24**	$0.1359 \div 3$

25	$0.0076 \div 4$	**29**	$6.3 \div 7$	**33**	$14.749 \div 7$
26	$0.81 \div 9$	**30**	$0.0636 \div 6$	**34**	$1.86 \div 3$
27	$0.5215 \div 5$	**31**	$0.038 \div 2$	**35**	$0.222 \div 6$
28	$0.000075 \div 5$	**32**	$4.62 \div 6$	**36**	$6.24 \div 8$

It may be necessary to add zeros at the end of a number in order to finish the division.

$2.9 \div 8$

$2.9 \div 8 = 0.3625$

$$\begin{array}{r} {\scriptstyle 524} \\ 8\overline{)2.9000} \\ 0.3625 \end{array}$$

Find the value of:

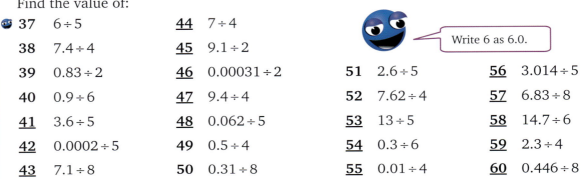

37	$6 \div 5$	**44**	$7 \div 4$				
38	$7.4 \div 4$	**45**	$9.1 \div 2$				Write 6 as 6.0.
39	$0.83 \div 2$	**46**	$0.00031 \div 2$	51	$2.6 \div 5$	**56**	$3.014 \div 5$
40	$0.9 \div 6$	**47**	$9.4 \div 4$	52	$7.62 \div 4$	**57**	$6.83 \div 8$
41	$3.6 \div 5$	**48**	$0.062 \div 5$	**53**	$13 \div 5$	**58**	$14.7 \div 6$
42	$0.0002 \div 5$	49	$0.5 \div 4$	54	$0.3 \div 6$	**59**	$2.3 \div 4$
43	$7.1 \div 8$	50	$0.31 \div 8$	55	$0.01 \div 4$	**60**	$0.446 \div 8$

If we divide 7.8 m of tape equally amongst 5 people, how long a piece will they each have?

We need to divide 7.8 m into 5 equal lengths, so we need to find $7.8 \div 5$.

Length of each piece $= 7.8 \div 5$ m
$= 1.56$ m

$$\begin{array}{r} {\scriptstyle 23} \\ 5\overline{)7.80} \\ 1.56 \end{array}$$

61 The perimeter of a square is 14.6 cm. What is the length of a side?

62 Divide 32.6 m into 8 equal parts.

63 Share 14.3 kg equally between 2 people.

64 The perimeter of a regular pentagon (a five-sided figure with all the sides equal) is 16 cm. What is the length of one side?

65 Share $36 equally amongst 8 people.

Long division

We can also use long division. The decimal point is used only in the original number and the answer, and not in the lines of working below these.

Find $2.56 \div 16$

$2.56 \div 16 = 0.16$

```
        0.16
    16)2.56
        1 6
        ‾‾‾‾
          96
          96
          ‾‾
```

$4.2 \div 25$

$4.2 \div 25 = 0.168$

```
        0.168
    25)4.200
        2 5
        ‾‾‾‾
        1 70
        1 50
        ‾‾‾‾
          200
          200
          ‾‾‾
```

Find the value of:

1	$26.4 \div 24$	**7**	$0.0615 \div 15$	**13**	$35.52 \div 111$	**19**	$20.79 \div 99$
2	$2.1 \div 14$	**8**	$0.864 \div 24$	**14**	$7.28 \div 28$	**20**	$0.014\,26 \div 20$
3	$1.56 \div 13$	**9**	$8.48 \div 16$	**15**	$1.296 \div 54$	**21**	$23.4 \div 45$
4	$9.45 \div 21$	**10**	$5.2 \div 20$	**16**	$0.008\,05 \div 35$	**22**	$71.76 \div 23$
5	$11.22 \div 22$	**11**	$7.84 \div 14$	**17**	$54.4 \div 17$	**23**	$39.48 \div 47$
6	$80 \div 25$	**12**	$25.2 \div 36$	**18**	$21.93 \div 51$	**24**	$0.2556 \div 45$

Changing fractions to decimals (exact values)

We may think of $\frac{3}{4}$ as $3 \div 4$ and hence write it as a decimal.

Exercise 7l

Express $\frac{3}{4}$ as a decimal.

$\frac{3}{4} = 3 \div 4 = 0.75$

$$4\overline{)3.00}$$
$$0.75$$

Express the following fractions as decimals:

1 $\frac{1}{4}$

2 $\frac{3}{8}$

3 $\frac{3}{5}$

4 $\frac{5}{16}$

5 $\frac{1}{25}$

6 $2\frac{4}{5}$

7 $\frac{5}{8}$

8 $\frac{7}{16}$

9 $\frac{3}{25}$

10 $\frac{1}{32}$

Change $\frac{4}{5}$ to a decimal then add 2.

Standard decimals and fractions

It is worthwhile knowing a few equivalent fractions and decimals.
For example

$\frac{1}{2} = 0.5$ $\qquad$ $\frac{1}{4} = 0.25$ $\qquad$ $\frac{3}{4} = 0.75$ $\qquad$ $\frac{1}{8} = 0.125$

Exercise 7m

Write the following decimals as fractions in their lowest terms, without any working, if possible.

(Notice that $\frac{2}{5} = \frac{4}{10} = 0.4$.)

1 0.2 $\qquad$ **3** 0.8 $\qquad$ **5** 0.6 $\qquad$ **7** 0.9

2 0.3 $\qquad$ **4** 0.75 $\qquad$ **6** 0.7 $\qquad$ **8** 0.05

Write down the following fractions as decimals:

9 $\frac{9}{10}$ $\qquad$ **11** $\frac{4}{5}$ $\qquad$ **13** $\frac{3}{100}$ $\qquad$ **15** $\frac{5}{8}$

10 $\frac{1}{4}$ $\qquad$ **12** $\frac{3}{8}$ $\qquad$ **14** $\frac{3}{4}$ $\qquad$ **16** $\frac{7}{100}$

Investigation

Not all fractions convert to exact decimals.

For example, $\frac{1}{3} = 0.333...$ and so on for as long as you want to.

1   Investigate which of the fractions $\frac{1}{2}, \frac{1}{3}, \frac{1}{4}, \frac{1}{5}, \frac{1}{6}, \frac{1}{7}, \frac{1}{8}, \frac{1}{9}$ convert to exact decimals.

2   Investigate what happens if you change the numerators of these fractions. (Keep to proper fractions.)

3   Is there a connection between the denominators of the fractions in parts **1** and **2** that convert to exact decimals? Explain your answer.

4   Investigate some fractions with a two-digit denominator.

   Do not try converting all proper fractions with a two-digit denominator to decimals, but see if you can find a rule about the denominators that will tell you if a fraction will convert to an exact decimal. Test your rule on some fractions.

Long method of multiplication

Exercise 7n

From Chapter 6 you know how to multiply fractions. Convert each decimal to a fraction, then multiply the fractions.

Calculate the following products:

1	0.3×0.02	**7**	4×0.06
2	0.1×0.1	**8**	0.4×0.0012
3	0.003×6	**9**	0.08×0.01
4	3×0.02	**10**	0.0003×0.002
5	0.001×0.3	**11**	0.9×0.02
6	0.4×0.0001	**12**	0.004×2

$$0.04 \times 0.2 = \frac{4}{100} \times \frac{2}{10}$$
$$= \frac{8}{1000} = 0.008$$

Short method of multiplication

In the examples above, if we add together the number of digits (including zeros) after the decimal points in the original two numbers, we get the number of digits after the point in the answer.

The number of digits after the point is called the number of decimal places. In the first example in Exercise **7n**, 0.3 has one decimal place, 0.02 has two decimal places and the answer, 0.006, has three decimal places, which is the sum of one and two.

We can use this fact to work out 0.3×0.02 without using fractions. Multiply 3 by 2 ignoring the points; count up the number of decimal places after the points and then put the point in the correct position in the answer, writing in zeros where necessary, i.e. $0.3 \times 0.02 = 0.006$.

Any zeros that come after the point must be included when counting the decimal places.

This flow chart summarises the process.

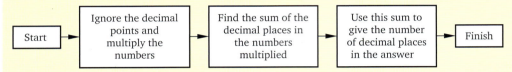

Exercise 7p

Find 0.08×0.4

First ignore the decimal points; just multiply the numbers together, i.e. $8 \times 4 = 32$. Now count the number of decimal places in each of the two numbers you are multiplying together. Adding them gives the number of places in the answer, counting back from the right-hand figure. Sometimes you have to put a 0 in too, because there aren't enough decimal places.

0.08	×	0.4	=	0.032	$8 \times 4 = 32$
(2 places)		(1 place)		(3 places)	

6×0.002

6	×	0.002	=	0.012	$6 \times 2 = 12$
(0 places)		(3 places)		(3 places)	

Calculate the following products:

1	0.6×0.3	**7**	0.5×0.07	**13**	0.07×12
2	0.04×0.06	**8**	8×0.6	**14**	4×0.009
3	0.009×2	**9**	0.08×0.08	**15**	0.9×9
4	0.07×0.008	**10**	3×0.0006	**16**	0.0008×11
5	0.12×0.09	**11**	0.7×0.06	**17**	7×0.011
6	0.07×0.0003	**12**	9×0.08	**18**	0.04×7

Zeros appearing in the multiplication in the middle or at the right-hand end must also be considered when counting the places.

0.252×0.4

0.252	×	0.4	=	0.1008	252
(3 places)		(1 place)		(4 places)	× 4
					1008

2.5×6

2.5	×	6	=	15.0	25
(1 place)		(0 places)		(1 place)	× 6
					150

300×0.2

300	×	0.2	=	60.0	$300 \times 2 = 600$
(0 places)		(1 place)		(1 place)	

Calculate the following products:

19	0.751×0.2	**27**	320×0.07	**35**	4×1.6
20	3.2×0.5	**28**	0.4×0.0055	**36**	5×0.016
21	0.35×4	**29**	0.5×0.06	**37**	0.00004×0.00016
22	1.52×0.0006	**30**	0.04×0.352	**38**	16000×0.05
23	400×0.6	**31**	1.6×0.4	**39**	0.16×4
24	31.5×2	**32**	1.6×0.5	**40**	0.0016×5
25	5.6×0.02	**33**	160×0.004	**41**	0.072×0.6
26	0.008×256	**34**	0.16×0.005	**42**	310×0.04

Multiplication of decimals

Exercise 7q

Find 0.26×1.3

0.26	×	1.3	=	0.338	26
(2 places)		(1 place)		(3 places)	× 13
					78
					260
					338

Calculate the following products:

1	4.2×1.6	**13**	14.4×4.5
2	52×0.24	**14**	1.36×0.082
3	0.68×0.14	**15**	0.081×0.032
4	48.2×26	**16**	1.6×1.6
5	310×1.4	**17**	0.16×16
6	1.68×0.27	**18**	0.0016×1600
7	13.2×2.5	**19**	0.28×0.28
8	0.0082×0.034	**20**	0.34×0.31
9	17.8×420	**21**	14×0.123
10	3.2×37	**22**	1.9×9.1
11	39×0.23	**23**	8.2×2.8
12	0.264×750	**24**	0.047×0.66

Problems

Exercise 7r

Find the cost of 6 books at $250.35 each.

The total cost is equal to the cost of 1 book multiplied by the number of books.

$\therefore$ Cost $= \$250.35 \times 6$
$= \$1502.10$

$$
\begin{array}{r}
25035 \\
\times \quad\quad 6 \\
\hline
150210
\end{array}
$$

1 Find the cost of 10 articles at $32.50 each.

2 The perimeter of a square is 17.6 cm. Find the length of one side of the square.

Read the question slowly to make sure you understand what you are being asked to do. Read it several times if necessary.

3 Divide 26.6 kg into 7 equal parts.

4 Find the perimeter of a square of side 4.2 cm.

5 Find the cost of 62 notebooks at $120.50 each.

6 Multiply 3.2 by 0.6 and divide the result by 8.

7 If 68.25 m of ribbon is divided into 21 equal pieces, how long is each piece?

<u>**8**</u>   The length of a side of a regular twelve-sided polygon
(a shape with 12 equal sides) is 4.2 m. Find the perimeter
of the polygon.

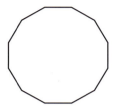

Mixed exercises

Exercise 7s

1   Write 0.02 as a fraction in its lowest terms.

2   Write $\frac{9}{1000}$ and $\frac{91}{1000}$ as decimals.

3   Add together 4.27, 31 and 1.6.

4   Subtract 1.82 from 4.2.

5   Divide 0.082 by 4.

6   Multiply 0.0301 by 100.

7   Express $\frac{7}{8}$ as a decimal.

8   Find the perimeter of the quadrilateral:

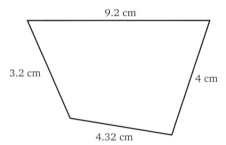

Exercise 7t

1   Give 0.3 as a fraction.

2   Express $\frac{14}{100}$ as a decimal.

3   Find the sum of 16.2, 4.12 and 7.

4   Find the value of $0.062 \div 100$.

5   Divide 1.5 by 25.

6   Subtract 14.8 from 16.3.

7   Find the total bill for three books costing $520.50, $610.50
and $240.50.

8 Which is bigger, $\frac{2}{5}$ or 0.3?

9 Multiply 1.9 by 2.5.

Exercise 7u

1 Give 0.008 as a fraction in its lowest terms.

2 Express $\frac{4}{5}$ as a decimal.

3 Add 14.2, 6, 0.38 and 7.21 together.

4 Subtract 14.96 from 100.

5 Divide 8.6 by 1000.

6 Evaluate $1.5 \div 6$.

7 Express $\frac{3}{16}$ as a decimal.

8 The perimeter of an equilateral triangle (all three sides are equal) is 14.4 cm. What is the length of one side?

9 Find 4.06×8.

Exercise 7v

1 Express $\frac{1}{8}$ as a decimal.

2 Find $8.2 - 1.92$.

3 Divide 1.3 by 5.

4 Add 4.2 and 0.28 and subtract 1.5 from the result.

5 Express 0.09 as a fraction.

6 Multiply 0.028 by ten thousand.

7 Divide 42 by 15.

8 I start with 19.44 m of rope. I cut off two pieces, one 7.39 m long and another 9.53 m long. How much do I have left?

9 Multiply 3.06 by 0.4.

In this chapter you have seen that...

✔ the decimal point divides the units from the tenths

✔ you can add and subtract decimals by writing them in columns, making sure that the decimal points are in line

✔ you can multiply decimals by 10, 100, ... by moving the digits to the left

✔ you can divide a decimal by a whole number by the same method you use for whole numbers provided you keep the decimal point in the answer above the decimal point in the number you are dividing into

✔ a fraction can be changed to a decimal by dividing the top of the fraction by the bottom

✔ a decimal can be changed to a fraction by writing the numbers after the point as tenths, hundredths, ... and simplifying

✔ you should learn that $\frac{1}{2} = 0.5$, $\frac{1}{4} = 0.25$, $\frac{3}{4} = 0.75$, $\frac{1}{8} = 0.125$

✔ the sum of the decimal places in the numbers that are multiplied gives the number of decimal places in the answer.

REVIEW TEST 1: CHAPTERS 1–7

In questions **1** to **13**, choose the letter for the correct answer.

1 To the nearest ten, 187 =

   A 100   **B** 180   **C** 190   **D** 200

2 Written as a fraction 1.4 =

   A $1\frac{4}{7}$ **B** $1\frac{4}{9}$ **C** $1\frac{4}{10}$ **D** $1\frac{4}{11}$

3 Given $X = \{2, 4, 6, 8\}$, $Y = \{4, 8, 12, 16\}$, then $X \cap Y =$

   A $\{4, 8\}$ **B** $\{2, 4, 6, 8, 12, 16\}$

   C $\{2, 6\}$ **D** $\{2, 6, 12, 16\}$

4 The LCM of 2, 4 and 5 is

   A 20   **B** 30   **C** 40   **D** 50

5 $2\frac{1}{3} \times 1\frac{2}{7} =$

   A $\frac{1}{21}$ **B** $1\frac{1}{21}$ **C** $2\frac{2}{21}$ **D** $\frac{3}{1}$

6 Written as a decimal, $\frac{9}{10} + \frac{7}{1000} =$

   A 0.097   **B** 0.907   **C** 0.97   **D** 9.07

7 $0.3 \times 0.02 =$

   A 0.006   **B** 0.060   **C** 0.600   **D** 6.000

8 The HCF of 12, 15 and 30 is

   A 3   **B** 6   **C** 60   **D** 180

9 All the factors of 12 are

   A 1, 2, 3, 4, 6, 12   **B** 2, 3, 4, 6

   C 2, 4, 6, 8   **D** 3, 6, 9, 12

10 $2 \times 1\frac{3}{4} =$

   A $2\frac{3}{4}$ **B** $2\frac{3}{8}$ **C** $3\frac{1}{2}$ **D** $3\frac{3}{4}$

11 $0.8 \div 0.4 =$

   A 0.02   **B** 0.2   **C** 0.32   **D** 2

12 The remainder when 70 is divided by 15 is

   A 5   **B** 10   **C** 15   **D** 60

13 $A = \{$square numbers less than 10$\}$ and $B = \{$prime numbers less than 10$\}$.
$A \cap B =$

A $\{1, 9\}$ B $\{9\}$ C $\{2\}$ D $\{\}$

14 **a** Simplify $\frac{7}{12} \div \frac{21}{4}$

b Write down the next two numbers in this pattern: 1, 3, 6, ...

c Calculate $3 \times 10 + 7 \times 10 \div (2 \times 10) + 0 \times 1$

15 **a** Find the greatest common divisor of 60 and 72.

b Given that $12\,740 = 2 \times 2 \times 5 \times 7 \times 7 \times 13$,

does 13 divide $12\,740$?   Why?

does 9 divide $12\,740$?   Why?

16 **a** An engine uses $\frac{3}{16}$ litres of oil in 60 minutes. How many litres of oil will it use in 190 minutes at this rate?

b If $\frac{7}{8}$ of a fence can be built in $3\frac{1}{2}$ hours, what fraction of the fence would be done in 1 hour working at the same rate?

17 **a** Given $P = \{$factors of 36$\}$, $Q = \{$factors of 42$\}$, list the members of

i $P \cap Q$ **ii** $P \cup Q$

b Show in a Venn diagram $A = \{2, 3, 4, 6, 12\}$, $B = \{2, 4, 8\}$.

18 **a** Arrange the following fractions in ascending order

$\frac{2}{3}, \frac{7}{9}, \frac{3}{4}, \frac{5}{12}$

b Find 8 months as a fraction of 2 years.

8 Metric units

Did you know?

The word zero came from the Italian. It was not always as important as it is today.

We were using numbers for thousands of years before zero (0) was introduced to us.

Zero is special:

- If we add or subtract 0 from a number, the result is the original number.
- If we multiply a number by 0, the result is zero.
- If we divide 0 by a number other than 0, the result is 0.
- Alas! We cannot define a number divided by 0.

Oh wonderful 0!

You need to know...

✔ how to multiply and divide by 10, 100 and 1000
✔ how to multiply by any number

Key words

centimetre, gram, kilogram, kilometre, mass, metre, millimetre, perimeter, rectangular, tonne, the symbol ∴

In times long ago people were totally self-contained within their local community. They bartered for what they needed. As time went on a coinage evolved. With coins you could buy what you wanted and you could receive coins for something you wanted to sell, including your labour.

Alongside the creation of coinage came the need to know how much of something you were buying or selling. It is difficult to compare quantities when different measures are used. One group might have used a sackful of wheat and another group may have used a potful. This need evolved into the standard units we have today.

Whenever we want to measure a length, or weigh an object, we find the length or mass in standard units. We might, for instance, give the length of a line in millimetres or the mass of a bag of apples in pounds. The millimetre belongs to a set of units called the metric system. The pound is one of the imperial units.

The metric system was developed in France in 1790 so that units in the system would be related to each other by a factor of ten.

Units of length

The basic unit of length is the *metre* (m). To get an idea of how long a metre is, remember that a standard bed is about 2 m long. However, a metre is not a useful unit for measuring either very large things or very small things so we need larger units and smaller units.

We get the larger unit by multiplying the metre by 1000. We get the smaller units by dividing the metre into 100 parts or 1000 parts.

1000 metres is called 1 kilometre   (km)

(It takes about 15 minutes to walk a distance of 1 km.)

$\frac{1}{100}$ of a metre is called 1 centimetre   (cm)

$\frac{1}{1000}$ of a metre is called 1 millimetre   (mm)

(You can see centimetres and millimetres on your ruler.)

Some uses of metric units of length

Millimetres (mm) for lengths of nails and screws, widths of film, tapes and ribbons.

Centimetres (cm) for body sizes, i.e. height, chest, etc, widths of wallpaper, belts, and ties.

Metres (m) for sizes of rooms, swimming pools, garden beds, hoses, ladders, etc.

Kilometres (km) for road signs, maps and distances between places.

Exercise 8a

1 Which metric unit would you use to measure
  a the length of your classroom
  b the length of your pencil
  c the length of a soccer pitch
  d the distance from Castries to Roseau
  e the length of a page in this book
  f the thickness of your exercise book?

2 Use your ruler to draw a line of length

a 10 cm	**b** 3 cm	**c** 15 cm	**d** 50 mm
e 20 mm	**f** 4 cm	**g** 15 mm	**h** 12 cm
i 25 mm	**j** 16 mm	**k** 5 cm	**l** 75 mm

3 Estimate the length, in centimetres, of the following lines:

  a _____

  b _____

  c _____

  d ___

  e _____

Use a ruler to measure its length. Put 0 on the ruler over one end. Now move your eye until it is over the mark on the ruler at the other end of the line. This value on the ruler gives the length of the line.

  Now use your ruler to measure each line.

4 Estimate the length, in millimetres, of the following lines:

  a _____

  b _____

  c ___

  d _____

  e _____

  Now use your ruler to measure each line.

5 Use a straight edge (not a ruler with a scale) to draw a line that is approximately
  a 10 cm long   **b** 5 cm long   **c** 15 cm long   **d** 20 mm long.
  Now measure each line to see how good your approximation was.

6   Estimate the width of your classroom in metres.

7   Estimate the length of your classroom in metres.

8   Measure the length and width of your exercise book in centimetres. Draw a rough sketch of your book with the measurements on it. Find the perimeter (the distance all round) of your book.

9   Each side of a square is 10 cm long. Draw a rough sketch of the square with the measurements on it. Calculate the perimeter of the square.

10   A sheet is 200 cm wide and 250 cm long. What is the perimeter of the sheet?

? Practical work

This shows a woman near a tree.

The woman is 170 cm tall.

a Estimate the height of the tree.

b Use a person or an object (e.g. a door) whose height you know to estimate the height and width of the main building in your school.

c Explain how you could estimate the length and height of a bridge.

Changing from large units to smaller units

The metric units of length are the kilometre, the metre, the centimetre and the millimetre where

$$1\,\text{km} = 1000\,\text{m} \qquad\qquad 1\,\text{m} = 100\,\text{cm}$$

$$1\,\text{m} = 1000\,\text{mm} \qquad\qquad 1\,\text{cm} = 10\,\text{mm}$$

Exercise 8b

Express 3 km in metres.

1 km is 1000 m, so 3 km is 3 times 1000 m

$$3\,\text{km} = 3 \times 1000\,\text{m}$$

$$= 3000\,\text{m}$$

131

Express 3.5 m in centimetres.

1 m is 100 cm, so 3.5 m is 3.5 times 100 cm

$$3.5 \, m = 3.5 \times 100 \, cm$$
$$= 350 \, cm$$

Express the given quantity in terms of the unit in brackets.

1	2 m	(cm)	9	3 m	(mm)	17	1.9 m	(mm)
2	5 km	(m)	10	2 km	(mm)	18	3.5 km	(m)
3	3 cm	(mm)	11	5 m	(cm)	19	2.7 m	(cm)
4	4 m	(cm)	12	7 m	(mm)	20	1.9 km	(cm)
5	12 km	(m)	13	1.5 m	(cm)	21	3.8 cm	(mm)
6	15 cm	(mm)	14	2.3 cm	(mm)	22	9.2 m	(mm)
7	6 m	(mm)	15	4.6 km	(m)	23	2.3 km	(m)
8	1 km	(cm)	16	3.7 m	(mm)	24	8.4 m	(cm)

Units of mass

The most familiar units used for weighing are the kilogram (kg) and the gram (g). We shall use the term 'mass', not 'weight'.

Most groceries that are sold in tins or packets have masses given in grams. For example, the mass of the most common packet of butter is 250 g. One eating apple weighs roughly 100 g, so the gram is a small unit of mass.

Kilograms are used to give the mass of sugar or flour: the mass of the most common bag of sugar is 1 kg and the most common bag of flour weighs 1.5 kg.

For weighing large loads (timber or steel, for example) a larger unit of mass is needed, and we use the tonne (t). For weighing very small quantities (the mass of a particular drug in one pill, for example) we use the milligram (mg).

The relationships between these masses are

$$1 \, t = 1000 \, kg$$
$$1 \, kg = 1000 \, g$$
$$1 \, g = 1000 \, mg$$

Exercise 8c

Express each quantity in terms of the unit given in brackets.

Express 2t in grams.

First change tonnes to kg, then change kg to grams.
1t is 1000 kg and 1 kg is 1000 g

$$2t = 2 \times 1000 \, kg$$
$$= 2000 \, kg$$
$$= 2000 \times 1000 \, g$$
$$= 2\,000\,000 \, g$$

1	12t	(kg)	**9**	4kg	(g)	**17**	5.2kg	(mg)
2	3kg	(g)	**10**	2kg	(mg)	**18**	0.6g	(mg)
3	5g	(mg)	**11**	3t	(kg)	**19**	11.3t	(kg)
4	1t	(g)	**12**	4g	(mg)	**20**	2.5kg	(g)
5	1kg	(mg)	**13**	1.5kg	(g)	**21**	7.3g	(mg)
6	13kg	(g)	**14**	2.7t	(kg)	**22**	0.3kg	(mg)
7	6g	(mg)	**15**	1.8g	(mg)	**23**	0.5t	(kg)
8	2t	(g)	**16**	0.7t	(kg)	**24**	0.8g	(mg)

Mixed units

When you use your ruler to measure a line, you will probably find that the line is not an exact number of centimetres. For example, the width of this page is 16 cm and 9 mm. We can say that the width of this page is 16 cm 9 mm or we could give the width in millimetres alone.

Now $$16 \, cm = 16 \times 10 \, mm$$
$$= 160 \, mm$$

So $$16 \, cm \, 9 \, mm = 169 \, mm$$

Exercise 8d

Express each quantity in terms of the unit given in brackets.

4 kg 50 g   (g)

Change 4 kg to grams, then add 50 g.

$$4 \, kg = 4 \times 1000 \, g$$
$$= 4000 \, g$$

Therefore $4 \, kg \, 50 \, g = 4050 \, g$

1	1 m 36 cm	(cm)		**11**	3 kg 500 g	(g)
2	3 cm 5 mm	(mm)		**12**	2 kg 8 g	(g)
3	1 km 50 m	(m)		**13**	5 g 500 mg	(mg)
4	4 cm 8 mm	(mm)		**14**	2 t 800 kg	(kg)
5	2 m 7 cm	(cm)		**15**	3 t 250 kg	(kg)
6	3 km 20 m	(m)		**16**	1 kg 20 g	(g)
7	5 m 2 cm	(cm)		**17**	1 g 250 mg	(mg)
8	5 km 500 m	(m)		**18**	3 kg 550 g	(g)
9	20 cm 2 mm	(mm)		**19**	2 t 50 kg	(kg)
10	8 m 9 mm	(mm)		**20**	1 kg 10 g	(g)

Changing from small units to larger units

Exercise 8e

Express 400 cm in metres.

100 cm = 1 m, so 1 cm = 1 ÷ 100 m

So   400 cm = 400 ÷ 100 m
                  = 4 m

In questions **1** to **20**, express the given quantity in terms of the unit given in brackets:

1	300 mm	(cm)		**4**	250 mm	(cm)
2	6000 m	(km)		**5**	1600 m	(km)
3	150 cm	(m)		**6**	72 m	(km)

7	12 cm	(m)		**14**	5020 g	(kg)
8	88 mm	(cm)		**15**	3800 kg	(t)
9	1250 mm	(m)		**16**	86 kg	(t)
10	2850 m	(km)		**17**	560 g	(kg)
11	1500 kg	(t)		**18**	28 mg	(g)
12	3680 g	(kg)		**19**	190 kg	(t)
13	1500 mg	(g)		**20**	86 g	(kg)

5 m 36 cm   (m)

First change 36 cm to metres, then add 5.

$$36 \text{ cm} = 36 \div 100 \text{ m}$$
$$= 0.36 \text{ m}$$

So $\qquad\qquad$ 5 m 36 cm = 5.36 m.

In questions **21** to **40** express the given quantity in terms of the unit given in brackets:

21	3 m 45 cm	(m)		**31**	5 kg 142 g	(kg)
22	8 cm 4 mm	(cm)		**32**	48 g 171 mg	(g)
23	11 km 2 m	(km)		**33**	9 kg 8 g	(kg)
24	2 km 42 m	(km)		**34**	9 g 88 mg	(g)
25	4 cm 4 mm	(cm)		**35**	12 kg 19 g	(kg)
26	5 m 3 cm	(m)		**36**	4 g 111 mg	(g)
27	7 km 5 m	(km)		**37**	1 t 56 kg	(t)
28	4 m 5 mm	(m)		**38**	5 g 3 mg	(g)
29	1 km 10 cm	(km)		**39**	250 g 500 mg	(kg)
30	8 cm 5 mm	(km)		**40**	850 kg 550 g	(t)

Adding and subtracting metric quantities

These diagrams summarise how to change between units.

This diagram shows the relationship between the main metric units of length. Remember that you multiply when you convert to a smaller unit and divide when you convert to a larger unit.

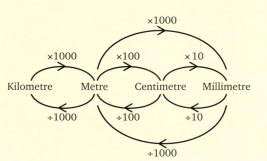

This diagram shows the relationship between the main metric units of mass.

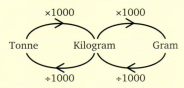

Exercise 8f

Find $1\,kg + 158\,g$ in   **a**   grams   **b**   kilograms.

a $1\,kg = 1000\,g$

$\therefore$ $1\,kg + 158\,g = 1158\,g$

($\therefore$ means 'therefore' or 'it follows that')

b $158\,g = 158 \div 1000\,kg$

                        $= 0.158\,kg$

$\therefore$ $1\,kg + 158\,g = 1.158\,kg$

Find the sum of $5\,m$, $4\,cm$ and $97\,mm$ in   **a**   metres   **b**   centimetres.

a $4\,cm = 4 \div 100\,m = 0.04\,m$

                $97\,mm = 97 \div 1000\,m = 0.097\,m$

$\therefore$ $5\,m + 4\,cm + 97\,mm = (5 + 0.04 + 0.097)\,m$

                        $= 5.137\,m$

b $5\,m = 5 \times 100\,cm = 500\,cm$

                $97\,mm = 97 \div 10\,cm = 9.7\,cm$

$\therefore$ $5\,m + 4\,cm + 97\,mm = (500 + 4 + 9.7)\,cm$

                        $= 513.7\,cm$

Alternatively, use your answer from **a**: $5.137\,m = 5.137 \times 100\,cm$

                        $= 513.7\,cm$

Quantities must be expressed in the same units before they are added or subtracted.

Find, giving your answer in metres:

1 $5\,m + 86\,cm$

2 $92\,cm + 115\,mm$

3 $3\,km + 136\,cm$

If you are changing to a smaller unit, e.g. from metres to centimetres, *multiply*.
If you are changing to a larger unit, e.g. from grams to kilograms, *divide*.

4   51 m + 3 km

5   36 cm + 87 mm + 520 cm

6   120 mm + 53 cm + 4 m

Find, giving your answer in millimetres:

7   36 cm + 80 mm

8   5 cm + 5 mm

9   1 m + 82 cm

10 2 m + 45 cm + 6 mm

11 3 cm + 5 cm + 2.9 cm

12 34 cm + 18 mm + 1 m

Find, giving your answer in grams:

13 3 kg + 250 g

14 5 kg + 115 g

15 5.8 kg + 9.3 kg

16 1 kg + 0.8 kg + 750 g

17 116 g + 0.93 kg + 680 mg

18 248 g + 0.06 kg + 730 mg

Find, expressing your answer in kilograms:

19 2 t + 580 kg

20 1.8 t + 562 kg

21 390 g + 1.83 kg

22 1.6 t + 3.9 kg + 2500 g

23 1.03 t + 9.6 kg + 0.05 t

24 5.4 t + 272 kg + 0.3 t

Find, expressing your answer in the unit given in brackets:

25 8 m − 52 cm   (cm)

26 52 mm + 87 cm (m)

27 1.3 kg − 150 g   (g)

28 1.3 m − 564 mm (cm)

29 2.05 t + 592 kg (kg)

30 20 g − 150 mg   (mg)

31 36 kg − 580 g   (g)

32 1.5 t − 590 kg   (kg)

33 3.9 m + 582 mm (cm)

34 0.3 m − 29.5 cm (mm)

Accuracy of measurements

a This is a map of an island.
Explain how you could estimate the length of its coastline.

b This is the same island, drawn to a larger scale.
Would you get the same answer for the length of its coastline
from this drawing?

c This shows the coast line of part of the island drawn to a much larger scale. If you used a map of the whole island drawn with this scale, how would your estimate of the length of the coast line compare with your first estimate?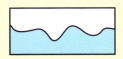

d Do you think it is possible to measure the length of the coastline exactly? (Think of a bit of coastline you know and imagine measuring a short length of it.)

e Now suppose that you want to measure the length of the table you are sitting at.

You could measure it with a ruler.

You could measure it with a tape measure marked in centimetres and millimetres.

You could measure it with a precision instrument that will read lengths to tenths of a millimetre, or even hundredths of a millimetre.

You could measure the length in several different places.
Write down, with reasons, whether it is possible to find the length exactly.

Do you think it is possible to give any measurement exactly?

Multiplying metric units

Exercise 8g

Calculate, expressing your answer in the unit given in brackets:

$3 \times 2\,\text{g}\ 741\,\text{mg}$ (g)

First express the mass in grams.

$2\,\text{g}\ 741\,\text{mg} = 2.741\,\text{g}$

$\therefore 3 \times 2\,\text{g}\ 741\,\text{mg} = 3 \times 2.741\,\text{g}$

$= 8.223\,\text{g}$

$$\begin{array}{r} 2741 \\ \times \quad 3 \\ \hline 8223 \end{array}$$

1 $4 \times 3\,\text{kg}\ 385\,\text{g}$ (g)

2 $9 \times 5\,\text{m}\ 88\,\text{mm}$ (mm)

3 $3 \times 4\,\text{kg}\ 521\,\text{g}$ (kg)

4 $5 \times 2\,\text{m}\ 51\,\text{cm}$ (m)

5 $10 \times 3\,\text{t}\ 200\,\text{kg}$ (t)

6 $2 \times 5\,\text{cm}\ 3\,\text{mm}$ (cm)

> First change the measurement to the unit required.

7 $6 \times 2\,\text{g}\ 561\,\text{mg}$ (mg)

8 $8 \times 3\,\text{km}\ 56\,\text{m}$ (km)

9 $3 \times 7\,\text{t}\ 590\,\text{kg}$ (t)

10 $7 \times 2\,\text{km}\ 320\,\text{m}$ (m)

Puzzle

If a box of bananas weighs 7 kilograms and half of its own mass,
how much does a box and a half of bananas weigh?

Problems

Exercise 8h

Find, in kilograms, the total mass of a bag of flour of mass 1.5 kg, a jar of
jam of mass 450 g and a packet of rice of mass 500 g.

The total mass means the sum of the three masses.
First change each mass to kg, then add them.

$$\text{The mass of the jar of jam} = 450 \div 1000 \, \text{kg}$$
$$= 0.45 \, \text{kg}$$

$$\text{The mass of the packet of rice} = 500 \div 1000 \, \text{kg}$$
$$= 0.5 \, \text{kg}$$

$$\text{The total mass} = (1.5 + 0.45 + 0.5) \, \text{kg}$$
$$= 2.45 \, \text{kg}$$

1 Find the sum, in metres, of 5 m, 52 cm, 420 cm.

2 Find the sum, in grams, of 1 kg, 260 g, 580 g.

3 Subtract 52 kg from 0.8 t, giving your answer in kilograms.

4 Find the difference, in grams, between 5 g and 890 mg.

5 Find the total length, in millimetres, of a piece of wood
   82 cm long and another piece of wood 260 mm long.

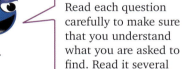

Read each question
carefully to make sure
that you understand
what you are asked to
find. Read it several
times if necessary.

6 Find the total mass, in kilograms, of 500 g of butter, 2 kg of potatoes,
   1.5 kg of flour.

7 One tin of baked beans has a mass of 220 g. What is the mass, in
   kilograms, of ten of these tins?

8 One fence post is 150 cm long. What length of wood, in metres, is
   needed to make ten such fence posts?

9 Find the perimeter of a square if each side is of length 8.3 cm. Give
   your answer in centimetres.

10 A wooden vegetable crate and its contents have a mass of 6.5 kg.
    If the crate has a mass of 1.2 kg what is the mass of its contents?

Time

Time is measured in millennia (1 millennium = 1000 years), centuries, decades, years, months, weeks, days, hours, minutes and seconds.

There are 12 months in a year but the number of days in a month varies.

There are 365 days in a year, except for leap years when there are 366.

Remember:
"Thirty days hath September, April, June and November. All the rest have thirty-one except for February clear which has twenty-eight and twenty-nine in each leap year."

The relationships between weeks, days, hours, minutes and seconds are fixed:

$$1 \text{ week} = 7 \text{ days}$$
$$1 \text{ day} = 24 \text{ hours}$$
$$1 \text{ hour} = 60 \text{ minutes}$$
$$1 \text{ minute} = 60 \text{ seconds}$$

When you change units of time, remember that you multiply when you change to a smaller unit, and you divide when you change to a larger unit.

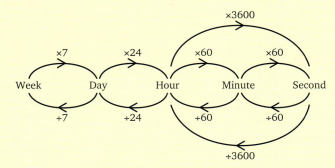

There are two ways of measuring the time of day: the 24-hour clock and the 12-hour clock.

The 24-hour clock uses the full 24 hours in a day, measuring from midnight through to the next midnight.

The time is given as a four figure number, for example, 1346 hr and 0730 hr.

The first two figures give the hours and the second two figures give the minutes.
So 1346 hr means 13 hours and 46 minutes after midnight and 0730 hr means 7 hours and 30 minutes after midnight.
Sometimes there is a space or a colon between the hours and the minutes, for example, 13 46 or 13:46.

The 12-hour clock uses the 12 hours from midnight to midday as a.m. times and the 12 hours from midday to midnight as p.m. times

a.m. is short for ante meridian and means before midday. p.m. is short for post meridian and means after midday.

The time is written as a number of hours and a number of minutes followed by a.m. or p.m.

The hours and the minutes are usually separated by a stop – for example, 6.30 a.m. means 6 hours and 30 minutes after midnight and 10.05 p.m. means 10 hours and 5 minutes after midday.

Midnight and midday are neither a.m. or p.m.

In the 24-hour clock, it is clear that 0000 hr means midnight and 1200 hr means midday.

But in the 12-hour clock, you need to write 'midnight' or 'midday' because 12.00 could mean either.

Noon is another word for midday.

The time on this clock can be read as 2.56 p.m. or 1456 hr.

Exercise 8i

1 Look at this calendar

Mon.	Tue.	Wed.	Thur.	Fri.	Sat.	Sun.
		1	2	3	4	5
6	7	8	9	10	11	12
13	14	15	16	17	18	19
20	21	22	23	24	25	26
27	28	29	30			

a Which month is this – August, September or October?
b Today is the 9th of the month. What day of the week is it?

c Today is the 24th of the month. What was the date a week ago today?

d The day after tomorrow is the third Wednesday of the month. What is the date today?

2 Elspeth goes on holiday on 8 June.

She returns on 21 June.

How many nights is she away?

3 David starts work on 1 September.

He gets paid on the twentieth of each month.

How many times does he get paid before Christmas?

4 The dates of birth of three people are:

    Julie 14/3/93   Dennis 14/1/92   Johanne 14/8/93

a Who is the eldest?

b Who is the youngest?

c In which year will the youngest be 30?

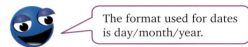

The format used for dates is day/month/year.

5 The president of the local cricket club is elected every year at the Annual General Meeting. This is a list of the presidents since the club was formed.

1908–22	S. Green
1922–28	P. Cave
1928–37	D. S. Short
1937–54	P. Baldrick
1954–62	H. Anthony
1962–76	D. S. Short
1976–79	W. May
1979–87	C. D. Bowen
1987–	O. D. Williams

a For how many years was P. Baldrick president?

b Who was president for the greatest number of years without a break?

c Assuming that O. D. Williams continued as president, in which year was he elected to begin his 25th year?

6 Write

a 190 minutes in hours and minutes.

b 450 hours in days and hours.

7 Write

a 5 minutes and 8 seconds in seconds.

b $3\frac{1}{2}$ hours in minutes.

8 Find

   a 20 minutes as a fraction of an hour

   b 36 seconds as a fraction of an hour.

9 These clock faces show the time at the beginning and end of a history lesson.

   a What time did the lesson start?

   b What time did the lesson end?

   c How long was the lesson?

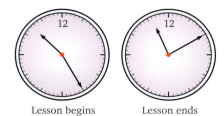

Lesson begins   Lesson ends

10 My ferry is due at 5.34 p.m.

   a How long should it be before it arrives?

   b Write 5.34 pm in 24-hour time.

11 Find the number of hours and minutes between

   a 9.30 a.m. and 11.15 a.m. the same day

   b 8.30 a.m. and 5.10 p.m. the same day

   c 10.20 p.m. and 12.30 a.m. the next day.

12 Mary sets the video recorder to start recording a programme at 3.50 p.m.

The programme lasts for $2\frac{1}{4}$ hours.

What time should she set the video to stop recording?

13 The time needed to cook a chicken is 40 minutes per kilogram plus 20 minutes.

How long should it take to cook a $3\frac{1}{2}$ kg chicken?

14 Susan's bus is due at 2005 hr.

2005 means 20 hours and 5 minutes after midnight.

   a How many minutes should she have to wait?

   b Write 2005 hr as an a.m. or p.m. time.

15 Find the period of time between
   a   0320 hours and 0950 hours on the same day
   b   0535 hours and 1404 hours on the same day
   c   2100 hours and 0500 hours next day
   d   0000 hours and 0303 hours next day.

Remember the time 0320 means 3 hours and 20 minutes after midnight. And 0305 means 3 hours and 5 minutes after midnight.

16 A plane leaves Kingston for Port of Spain.
   The flight should take 2 hours 10 minutes.
   The plane leaves Kingston on time at 1450 hours and is 15 minutes late arriving in Port of Spain. When does the plane arrive in Port of Spain?

17 The bus service from Westwick to Plimpton runs twice a day. This is the timetable.

Westwick	0945	1420
Red Farm Hill	1004	1439
Astleton arr.	1056	1531
dep.	1116	1545
Morgan's Hollow	1129	1559
Plimpton	1207	1637

   a   How long does each bus take to go from Westwick to Plimpton?
   b   Which two bus stops do you think are closest together?
       Give a reason for your answer.

Temperature

There are two commonly used units for measuring temperature.
One is degrees Celsius.
The freezing point of water is zero degrees Celsius. This is written 0°C.
The boiling point of water is 100 degrees Celsius. This is written 100°C.

The other is degrees Fahrenheit.
The freezing point of water is 32 degrees Fahrenheit. This is written 32°F.
The boiling point of water is 212°F.

Some thermometers have both scales on them.

Exercise 8j

1 **a** What is the temperature shown on this thermometer?

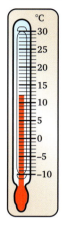

b What is the temperature shown on this thermometer?

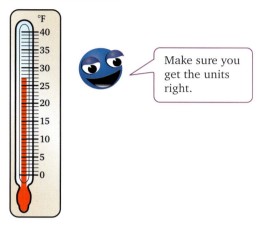

Make sure you get the units right.

c Which thermometer shows the higher temperature?

Give a reason for your answer.

You do not have to convert between Celsius and Fahrenheit to answer this.

2 This thermometer is marked in degrees Fahrenheit and in degrees Celsius.

a What Fahrenheit temperature does the thermometer show?

b What Celsius temperature does the thermometer show?

The temperature goes down by 20°C.

c What is the new Celsius reading?

d What is the new Fahrenheit reading?

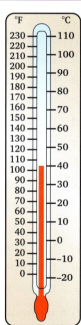

3 Use the thermometer in question **2** to convert

a 20°C to degrees Fahrenheit

b 5°C to degrees Fahrenheit

c 80°F to degrees Celsius

d 35°F to degrees Celsius.

4 In August 2003, the temperature in London reached a record high of 101°F.

Use these instructions to convert 101°F to degrees Celsius.

1 Subtract 32°.

2 Divide your answer by 9.

3 Multiply your answer by 5.

Check your answer on the thermometer in question 2.

Mixed problems

Exercise 8k

A girl takes to school a bag containing books, a shoe bag and a clarinet.

The contents of a bag have a mass of 5.3 kg. The shoe bag has a mass of 900 g and the clarinet has a mass of 1 kg 900 g. What is the mass, in kilograms, of the books?

The mass of the books is the difference between the combined mass of the shoe bag and clarinet and the mass of the bag.

First find the combined mass of the shoe bag and clarinet:
The shoe bag has a mass of 900 g = 0.9 kg
The clarinet has a mass of 1 kg 900 g = 1.9 kg
Together their mass is 0.9 kg + 1.9 kg = 2.8 kg

$$\begin{array}{r} 5.3 \\ -2.8 \\ \hline 2.5 \end{array}$$

The difference between 5.3 kg and 2.8 kg is
5.3 kg − 2.8 kg = 2.5 kg
So the books have a mass of 2.5 kg.

1 Write down the unit you would use to measure
   a   the mass of a fruit bun
   b   the time of a 100 m race
   c   the length of the room you sleep in
   d   the distance between Kingston and New York
   e   your height
   f   the temperature of the sea
   g   the mass of a bus
   h   the length of your foot
   i   the length of an eyelash
   j   the time it would take for a rocket to travel to Mars
   k   the time it takes to walk across the room you are in
   l   the mass of a hand of bananas.

2 A rectangular sheet of paper measures 32 cm by 17 cm. What is its perimeter
   a   in centimetres   b   in millimetres?

3 A girl travels to school by walking 450 m to the bus stop and then travelling 1 km 650 m by bus. The distance she walks after getting off the bus is 130 m. What distance is her total journey in kilometres?

4   A rectangular field is 947 m long and 581 m wide. What is the perimeter of the field? How many metres of fencing would be needed to go round the field leaving space for two gates each 3 m wide?

5   A man takes three parcels to the Post Office and has them weighed. One parcel has mass 4 kg 37 g, the mass of the second is 3 kg 982 g and the third one has mass 1 kg 173 g. What is their total mass in kilograms?

6   Wood is sometimes sold by the 'metric foot'. A metric foot is 30 cm. A man buys a length of wood which is 12 metric feet long. How long is the piece of wood in metres?

7   A freight train has five trucks. Two of them are carrying 15 t 880 kg each. Another has a load of 14 t 700 kg and the last two are each loaded with 24 t 600 kg. What is the total mass, in tonnes, of the contents of the five trucks? If the mass of each truck is 5 t 260 kg, what is the combined mass of the trucks and their contents?

8   A boy delivers newspapers by bicycle. The mass of the bicycle is 15.8 kg and the boy has mass 51.3 kg. At the beginning of the round the newspapers have mass 9.8 kg. What is the total mass of the boy and his bicycle loaded with newspapers? What is the mass when he has delivered half the newspapers?

9   Along a certain route Annotto Bay is 14.8 km from Buff Bay and Port Maria Bay is 24.6 km further on from Annotto Bay. If a car goes from Buff Bay to Annotto Bay then to Port Maria, and finally back to Buff Bay, how many kilometres has it travelled? At the beginning of the journey the car had enough petrol to go 80 km. At the end of its journey, how much further could it go before running out of petrol?

10 The instructions for converting a temperature of C degrees Celsius to F degrees Fahrenheit are multiply the Celsius temperature by 9 and divide the answer by 5. Then add 32.
    Use these instructions to convert 45°C to degrees Fahrenheit.

11 A flight from Amsterdam to Nice takes 2 hours 10 minutes.
The plane leaves Amsterdam at 1655.
Find the time the plane is due to arrive.

12 Passengers from a cruise ship berthed at Montego Bay are going to
Rafters' Village on a scheduled 5-hour excursion. The coach takes
1 hour 9 minutes to get to the village, they take a 92-minute rafting
trip on the Martha Brae, are given 35 minutes for shopping and it
take 95 minutes for the coach-ride back to the ship.
   a If the excursion began at 8.15 a.m., what time did they get back
   to the ship?
   b Lunch was arranged for 1.30 p.m. How long did they have to
   spare?

Mixed exercises

Exercise 8l

Express the given quantity in terms of the unit given in brackets:

1	4 km	(m)	**4**	250 g	(kg)	**7**	1 m 50 cm	(m)
2	30 g	(kg)	**5**	0.03 km	(cm)	**8**	2.8 cm	(mm)
3	3.5 m	(cm)	**6**	1250 m	(km)	**9**	65 g	(kg)

10 A tin of meat has mass 429 g. What is the mass, in kilograms, of ten
such tins?

Exercise 8m

Express the given quantity in terms of the unit in brackets:

1	236 cm	(m)	**5**	4 km 250 m	(km)
2	0.02 m	(mm)	**6**	3.6 t	(kg)
3	5 kg	(g)	**7**	2 kg 350 g	(kg)
4	500 mg	(g)	**8**	2 g	(mg)

9 Each side of a square is 65 cm long. What is the perimeter of the
square, in metres?

Exercise 8n

Express the given quantity in terms of the unit in brackets:

1   5.78 t   (kg)

2   3.54 m   (cm)

3   350 kg   (t)

4   0.155 mm   (cm)

5   1 t 560 kg   (t)

6   780 cm   (m)

7   $1\frac{1}{2}$ hours   (minutes)

8   2 km 50 m   (km)

9   A bus has mass 5 t 430 kg and carries 44 passengers each of whom is assumed to have a mass of 72 kg. Find the mass, in tonnes, of the bus and passengers when it is fully loaded.

Exercise 8p

Express the given quantity in terms of the unit in brackets:

1   4 cm 2 mm   (cm)

2   350 g   (kg)

3   1.52 kg   (g)

4   283 m   (km)

5   36 mm   (cm)

6   0.47 m   (mm)

7   36 cm   (m)

8   72 hours   (days)

9   A bag containing 5 c coins has mass 2.492 kg. If the mass of one coin is 7.12 g, how many coins are there in the bag?

In this chapter you have seen that...

✔ the metric units of length in common use are the kilometre, the metre, the centimetre and the millimetre, where

$$1\,cm = 10\,mm$$
$$1\,m = 100\,cm$$
$$1\,km = 1000\,m$$

✔ the metric units of mass in common use are the tonne, the kilogram, the gram and the milligram, where

$$1\,g = 1000\,mg$$
$$1\,kg = 1000\,g$$
$$1\,t = 1000\,kg$$

✔ the time of day can be measured as a.m. or p.m. times or as 24-hour time

✔ to change to a smaller unit, e.g. km to m, multiply

✔ to change to a larger unit, e.g. mm to cm, divide

✔ temperature is usually measured in degrees Celsius or degrees Fahrenheit, and that you can use a scale to convert from one to the other.

9 Imperial units

You need to know...

✔ the basic number facts including your tables

✔ how to deal with simple decimals

✔ the metric units of length and mass

Key words

approximation, foot, hundredweight, inch, kilometre, mass, metre, mile, ounce, pound, ton, tonne, yard

Units of length

Imperial units are still used in the USA. For instance, distances on road signs are still given in miles. One mile is roughly equivalent to $1\frac{1}{2}$ km. A better approximation is

> 5 miles is about 8 kilometres

Yards, feet and inches are other imperial units of length that are still used. In this system units are not always divided into ten parts to give smaller units so we have to learn 'tables'.

> 12 inches (in) = 1 foot (ft)
> 3 feet = 1 yard (yd)
> 1760 yards = 1 mile

Exercise 9a

Express 2 ft 5 in in inches.

First convert the number of feet into inches then add the odd number of inches.

$$2\,\text{ft} = 2 \times 12\,\text{in}$$
$$= 24\,\text{in}$$
$$\therefore \quad 2\,\text{ft}\,5\,\text{in} = 24 + 5\,\text{in}$$
$$= 29\,\text{in}$$

Express the given quantity in the unit in brackets:

1	5 ft 8 in	(in)	6	2 miles 800 yd	(yd)
2	4 yd 2 ft	(ft)	7	5 yd 2 ft	(ft)
3	1 mile 49 yd	(yd)	8	10 ft 3 in	(in)
4	2 ft 11 in	(in)	9	9 yd 1 ft	(ft)
5	8 ft 4 in	(in)	10	9 ft 10 in	(in)

52 in = (ft and in)

There are 12 inches in 1 foot so we need to find how many complete 12s there are in 52. A number of inches may be left over.

$$52\,\text{in} = 52 \div 12$$
$$= 4\,\text{ft}\,4\,\text{in}$$

$$\begin{array}{r} 4\ \text{r}4 \\ 12\overline{)52} \end{array}$$

11	36 in	(ft)	16	2000 yd	(miles and yd)
12	29 in	(ft and in)	17	75 in	(ft and in)
13	86 in	(ft and in)	18	100 ft	(yd and ft)
14	9 ft	(yd)	19	120 in	(ft and in)
15	13 ft	(yd and ft)	20	30 000 yd	(miles and yd)

Did you know?

Did you know that the unit of length used in horseracing is the furlong? This is a shortened form of 'furrow long' which was a convenient distance for a horse or ox to pull a plough before turning around to go back.

Units of mass

The imperial units of mass that are still used are pounds and ounces. Other units of mass that you may still see are hundredweights and tons (not to be confused with tonnes).

$$16 \text{ ounces (oz)} = 1 \text{ pound (lb)}$$
$$112 \text{ pounds} = 1 \text{ hundredweight (cwt)}$$
$$20 \text{ hundredweight} = 1 \text{ ton}$$

Exercise 9b

Express the given quantity in terms of the units given in brackets:

1	2 lb 6 oz	(oz)	**6**	24 oz	(lb and oz)
2	1 lb 12 oz	(oz)	**7**	18 oz	(lb and oz)
3	4 lb 3 oz	(oz)	**8**	36 oz	(lb and oz)
4	3 tons 4 cwt	(cwt)	**9**	30 cwt	(tons and cwt)
5	1 cwt 50 lb	(lb)	**10**	120 lb	(cwt and lb)

 Investigation

1 Some imperial units have specialised uses, for example, furlongs are used to measure distances in horse racing and fathoms are used to measure the depth of water.

a Use an internet search engine to find out the relationships between these units and the more common imperial units of length.

b Find out as much as you can about other imperial units of distance and mass.

c Nautical miles are used to measure distances at sea. Find out what you can about nautical miles, including the rough equivalence of 1 nautical mile in miles and in kilometres.

2 A group of young secondary school pupils were asked to write down their heights and masses on sheets of paper, which were gathered in. This is a list of *exactly* what was written down.

Height	Mass
141 cm	35 kg
1.38 cm	4 stone
1.8 m	6.26 stone
4 feet 5 inches	4 kg

Height	Mass
52 feet	6 stone
5 foot 4	8 stone
1 metre 53	$7\frac{1}{2}$ stone
1 metre 41 cm	28.0 kg
141 cm	5 stone 4 pounds
4 feet 7 inches	32 kg

a   This group of children used a mixture of units. Some of the entries are unbelievable.
   Which are they?
   Give some of the reasons for these unbelievable entries.

b   Find out how your group know their heights and masses; each of you write down your own height and mass on a piece of paper. Use whatever unit you know them in, and do not write your name on it. Collect in the pieces of paper and write out a list like the one above.

c   What official forms do you know about that ask for height? What unit is required?

d   Write down your own height and mass in both metric and imperial units.

Rough equivalence between metric and imperial units

If you shop in the USA you will find that goods are sold in pounds and ounces. It is often useful to be able to convert, roughly, pounds into kilograms or grams into pounds. For a rough conversion it is good enough to say that

   1 kg is about 2 lb

although one kilogram is slightly more than two pounds.

One metre is slightly longer than one yard but for a rough conversion it is good enough to say that

   1 m is about 1 yd

Remember that the symbol ≈ means 'is approximately equal to' so

   $1\,\text{kg} \approx 2\,\text{lb}$
   $1\,\text{m} \approx 1\,\text{yd or } 3\,\text{ft}$

Exercise 9c

In questions **1** to **10**, write the first unit roughly in terms of the unit in brackets:

> 5 kg   (lb)
> 1 kg ≈ 2 lb, so 5 kg is approximately 5 times 2 lb
> $$5\,kg \approx 5 \times 2\,lb$$
> $$\therefore \quad 5\,kg \approx 10\,lb$$
>
> 10 ft   (m)
> 3 ft ≈ 1 m, so you need to find the number of 3 s in 10
> $$10\,ft \approx 10 \div 3\,m$$
> $$\therefore \quad 10\,ft \approx 3.3\,m \text{ (to 1 d.p.)}$$

1	3 kg (lb)	3	4 lb (kg)	5	1.5 kg (lb)	7	3.5 kg (lb)	9	250 g (oz)
2	2 m (ft)	4	9 ft (m)	6	5 m (ft)	8	8 ft (m)	10	500 g (lb)

In questions **11** to **16** use the approximation 5 miles ≈ 8 km to convert the given number of miles into an approximate number of kilometres:

11	10 miles	13	15 miles	15	75 miles
12	20 miles	14	100 miles	16	40 miles

17   I buy a 5 lb bag of potatoes and two 1.5 kg bags of flour. What mass, roughly, in pounds do I have to carry?

18   A window is 6 ft high. Roughly, what is its height in metres?

19   I have a picture which measures 2 ft by 1 ft. Wood for framing it is sold by the metre. Roughly, what length of framing, in metres, should I buy?

20   Which is heavier, a 4 kg packet of sugar or a 5 lb bag of potatoes?

21   The distance between Antigua and St Kitts is about 50 miles. The distance between Dominica and Martinique is about 140 kilometres. What is the difference, in miles, between the distances the two pairs of islands are apart?

22   A recipe requires 250 grams of flour. Roughly, how many ounces is this?

Converting from inches to centimetres and from centimetres to inches is often useful. For most purposes it is good enough to say that 1 inch ≈ $2\frac{1}{2}$ cm.

23   An instruction in an old knitting pattern says knit 6 inches. Mary has a tape measure marked only in centimetres. How many centimetres should she knit?

24 The instructions for repotting a plant say that it should go into a 10 cm pot. The flower pots that Tom has in his shed are marked 3 in, 4 in and 5 in. Which one should he use?

25 Peter Stuart wishes to extend his gas pipe lines that were installed several years ago in 1 in and $\frac{1}{2}$-in diameter copper tubing. The only new piping he can buy has diameters of 10 mm, 15 mm, 20 mm or 25 mm. Use the approximation 1 in ≈ 2.5 cm to determine which piping he should buy that would be nearest to

    a the 1 inch pipes   **b** the $\frac{1}{2}$ inch pipes.

26 A carpenter wishes to replace a 6 in floorboard. The only sizes available are metric and have widths of 12 cm, 15 cm, 18 cm and 20 cm. Use the approximation 1 in ≈ 2.5 cm to determine which one he should buy.

27 A shop sells material at $10.50 per metre while the same material is sold in the local market at $9 per yard. Using 4 in ≈ 10 cm, find which is cheaper.

Did you know?

Why are there 112 pounds in a hundredweight?

Years ago, when a farmer had to pay tithes (a tithe is a tenth part) he had to pay the church one tenth of what he produced. One tenth of 112 is 11.2. Take this from 112 and you're left with 100.8. Rounded down to the nearest whole number this is 100. So, to have 100 pounds of wheat to sell a farmer needed to bring 112 pounds from the field. This is why there are 112 pounds in a hundredweight.

In this chapter you have seen that...

✔ common imperial units of length are inches (in), feet (ft), yards (yd), and miles and the relationships between them are:
12 in = 1 ft, 3 ft = 1 yd, 1760 yd = 1 mile

✔ common imperial units of mass are ounces (oz), pounds (lb), hundredweights (cwt) and tons and the relationships between them are:
16 oz = 1 lb, 112 lb = 1 cwt, 20 cwt = 1 ton

✔ you can roughly convert between metric and imperial units using
1 kg ≈ 2 lb, 1 m ≈ 1 yd, 5 miles ≈ 8 km

10 Introducing geometry

At the end of this chapter you should be able to...

1 Identify a line, a ray and a line segment.

2 Express the amount of 'turn' of a clock hand as a fraction of a revolution.

3 Express a change in direction as a fraction of a revolution.

4 Identify an angle as a change in direction.

5 Describe a right angle as a quarter of a revolution.

6 Define acute, obtuse and reflex angles in terms of right angles.

7 Identify acute, obtuse and reflex angles.

8 Define a degree as a fraction of a revolution.

9 Use a protractor to measure angles.

10 Draw angles of given size using a protractor.

11 State the properties of angles on a straight line.

You need to know...

✔ what an analogue clock face looks like

✔ the four main compass directions

✔ how to add and subtract whole numbers

✔ how to find a fraction of a quantity

Key words

acute angle, angles on a straight line, anticlockwise, bearing, clockwise, degree, line segment, obtuse, parallel, plane surface, perpendicular, protractor, rays, reflex, revolution, right angle, vertex

Basic concepts

Geometry is the mathematics of the properties of shapes. The building blocks of shapes are points and lines.

A point marks a position on a surface or in space.
A point has no width or length.
We mark points with a dot and use letters to identify different points.
In this diagram, C is the point on the vertex (corner) of the diagram and D is a point inside the diagram.

A line has no width but it does have length.
A line has no ends; it goes on for ever.
A line can be straight or curved.
We assume that a line is straight unless we are told it is curved.

A ray is a line with one end.

A line segment has two ends.

A curved line can form a closed loop such as:

A circle is a closed loop enclosing a plane surface where every point on the loop is the same distance from the centre. A plane surface means a perfectly flat surface.

Straight line

Curved line

Ray

Line segment

Centre

Circle

The distance round the circle is called the circumference.
A straight line segment from the centre to the circle is called a radius.
A straight line segment from one side of the circle to the other that goes through the centre is called a diameter.
A straight line segment from one side of the circle to the other that does not go through the centre is called a chord.

Circumference
Radius
Diameter
Chord

Polygons

A polygon is a plane shape bounded by straight line segments.

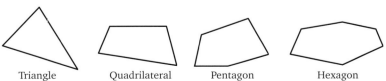

Triangle   Quadrilateral   Pentagon   Hexagon

The examples above are irregular polygons.

A regular polygon has all its sides equal and all its angles equal.
The diagrams below show regular polygons.

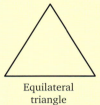
Equilateral
triangle

Square

Regular
hexagon

A rectangle is a quadrilateral, but it is not usually a regular
quadrilateral (i.e. a square).

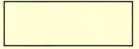

Exercise 10a

1 Each line in your exercise book is a line segment.
List three other line segments that you can see.

2 Match the correct description to each diagram.

a   **b**   **c**   **d**

  1. a straight line   2. a ray   3. a curved ray   4. a line segment

The arrows on the ends of the lines above show that they have no ends.
We do not usually show these arrows. So, in the diagrams below, we assume
that the lines go on forever unless they have an obvious end.

3 C and D are two points. Describe the line in each diagram.

a    **b**    **c**

4 Name each shape and state whether it is a regular polygon.

a   **b**   **c**

d   **e**

Fractions of a revolution

When the seconds hand of a clock starts at 12 and moves round until it stops at 12 again it has gone through one complete turn.

One complete turn is called a revolution.

When the seconds hand starts at 12 and stops at 3 it has turned through $\frac{1}{4}$ of a revolution.

Exercise 10b

What fraction of a revolution does the seconds hand of a clock turn through when:

1   it starts at 12 and stops at 9
2   it starts at 12 and stops at 6
3   it starts at 6 and stops at 9
4   it starts at 3 and stops at 9
5   it starts at 9 and stops at 12
6   it starts at 1 and stops at 7
7   it starts at 5 and stops at 11
8   it starts at 10 and stops at 4
9   it starts at 8 and stops at 8
10 it starts at 8 and stops at 11

11 it starts at 10 and stops at 2
12 it starts at 12 and stops at 4
13 it starts at 8 and stops at 5
14 it starts at 5 and stops at 2
15 it starts at 9 and stops at 5?

Draw a clock face with the initial and final positions of the hand, like the diagram above. Draw a curved arrow going clockwise from the start position to the end position of the hand.

Where does the seconds hand stop if:

16 it starts at 12 and turns through $\frac{1}{2}$ a turn
17 it starts at 12 and turns through $\frac{3}{4}$ of a turn
18 it starts at 6 and turns through $\frac{1}{4}$ of a turn
19 it starts at 9 and turns through $\frac{1}{2}$ of a turn
20 it starts at 6 and turns through a complete turn
21 it starts at 9 and turns through $\frac{3}{4}$ of a turn
22 it starts at 12 and turns through $\frac{1}{3}$ of a turn
23 it starts at 12 and turns through $\frac{2}{3}$ of a turn
24 it starts at 9 and turns through a complete turn
25 it starts at 6 and turns through $\frac{1}{2}$ a turn?

Bearings

The four main compass directions are north, south, east and west.

If you stand facing north and turn clockwise through $\frac{1}{2}$ a revolution you are then facing south.

Exercise 10c

1   If you stand facing west and turn anticlockwise through $\frac{3}{4}$ of a revolution, in which direction are you facing?

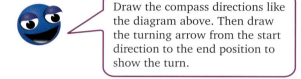

Draw the compass directions like the diagram above. Then draw the turning arrow from the start direction to the end position to show the turn.

2   If you stand facing south and turn clockwise through $\frac{1}{4}$ of a revolution, in which direction are you facing?

3   If you stand facing north and turn, in either direction, through a complete revolution, in which direction are you facing?

4   If you stand facing west and turn through $\frac{1}{2}$ a revolution, in which direction are you facing? Does it matter if you turn clockwise or anticlockwise?

5   If you stand facing south and turn through $1\frac{1}{2}$ revolutions, in which direction are you facing?

6   If you stand facing west and turn clockwise to face south what part of a revolution have you turned through?

7   If you stand facing north and turn clockwise to face west how much of a revolution have you turned through?

8   If you stand facing east and turn to face west what part of a revolution have you turned through?

Angles

When the hand of a clock moves from one position to another it has turned through an angle.

Right angles

A quarter of a revolution is called a *right angle*.

Half a revolution is two right angles.

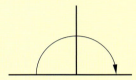

Exercise 10d

How many right angles does the seconds hand of a clock turn through when:

it starts at 3 and stops at 12?

Draw the start and end position of the hand with
a curved arrow from the start to the end position.

Now you can see that it turns through three right angles.

1	it starts at 6 and stops at 9	**5**	it starts at 12 and stops at 12
2	it starts at 3 and stops at 9	**6**	it starts at 8 and stops at 2
3	it starts at 12 and stops at 9	**7**	it starts at 9 and stops at 6
4	it starts at 3 and stops at 6	**8**	it starts at 7 and stops at 7?

How many right angles do you turn through if you:

9 face north and turn clockwise to face south

10 face west and turn clockwise to face north

11 face south and turn clockwise to face west

12 face north and turn anticlockwise to face east

13 face north and turn to face north again?

Acute, obtuse and reflex angles

Any angle that is smaller than a right angle is called an *acute angle*.

Any angle that is greater than one right angle and less than two right angles is called an *obtuse angle*.

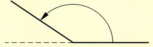

Any angle that is greater than two right angles is called a *reflex angle*.

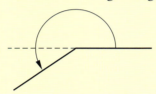

Exercise 10e

What type of angle is each of the following?

1

2

3

4

5

6

7

8

9

10

11

12

13

14

15

Degrees

One complete revolution is divided into 360 parts. Each part is called a *degree*. 360 degrees is written 360°.

360 seems a strange number of parts to have in a revolution but it is a good number because so many whole numbers divide into it exactly. This means that there are many fractions of a revolution that can be expressed as an exact number of degrees.

Exercise 10f

1 How many degrees are there in half a revolution?

2 How many degrees are there in one right angle?

3 How many degrees are there in three right angles?

How many degrees has the seconds hand of a clock turned through when it moves from 6 to 9?

Drawing the clockface as before shows that it has turned through 90°, i.e. 1 right angle.

How many degrees has the seconds hand of a clock turned through when it moves from:

4	12 to 6	**7**	9 to 3	**10**	7 to 11	**13**	4 to 10
5	3 to 6	**8**	9 to 6	**11**	1 to 10	**14**	5 to 8
6	6 to 3	**9**	2 to 5	**12**	8 to 5	**15**	6 to 12?

How many degrees has the second hand of a clock turned through when it moves from 6 to 8?

Drawing the clockface shows that the hand moves through 2 out 3 equal divisions of 90°, i.e. it moves through $\frac{2}{3}$ of 90° and $\frac{2}{3}$ of 90° $= \frac{2}{3} \times \frac{90°}{1} = 60°$.

Another way of looking at it is to say that the hand moves through 2 out of 12 equal divisions of a revolution, i.e. $\frac{2}{12}$ of 360°.

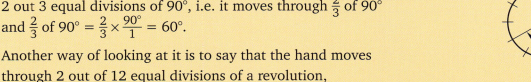

How many degrees has the seconds hand of a clock turned through when it moves from:

16	8 to 9	**26**	3 to 10
17	10 to halfway between 11 and 12	**27**	2 to 8
18	6 to 10	**28**	10 to 8
19	1 to 3	**29**	12 to 11
20	3 to halfway between 4 and 5	**30**	9 to 2
21	4 to 5	**31**	8 to 3
22	7 to 11	**32**	7 to 5
23	5 to 6	**33**	10 to 5
24	7 to 9	**34**	11 to 4
25	11 to 3	**35**	2 to 9?

 Investigation

The Babylonians chose to divide one complete revolution into 360 degrees. We still use this division. Why do you think that this division has not been decimalised?

You may find that the answer is clear if you list all the numbers between 1 and 20 that divide exactly into 360 and that divide exactly into 100.

Using a protractor to measure angles

A protractor looks like this:

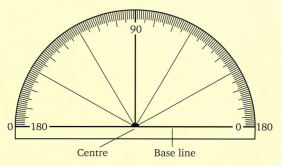

It has a straight line at or near the straight edge. This line is called the *base line*.

The *centre* of the base line is marked.

The protractor has two scales, an inside one and an outside one.

To measure the size of this angle, first decide whether it is acute or obtuse.

This is an acute angle because it is *less* than 90°.

Next place the protractor on the angle as shown.

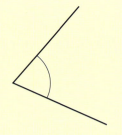

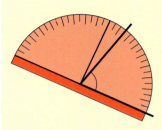

One arm of the angle is on the base line.

The vertex (point) of the angle is at the centre of the base line.

Choose the scale that starts at 0° on the arm of the base line. Read off the number where the other arm cuts this scale.

Check with your estimate to make sure that you have chosen the right scale.

Exercise 10g

Measure the following angles (if necessary, turn the page to a convenient position):

1
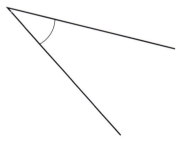

3
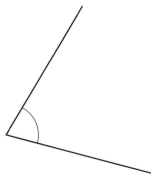

2
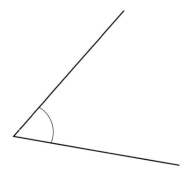

4
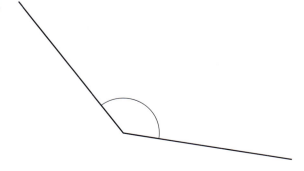

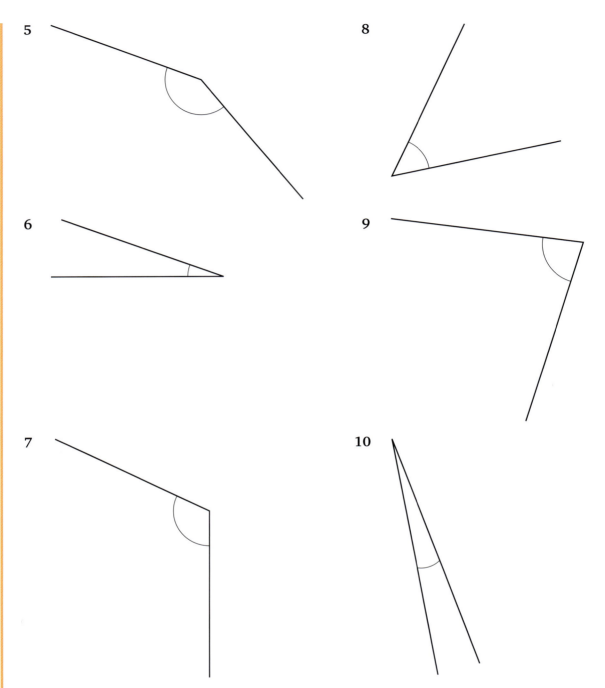

5

8

6

9

7

10

In questions **11** to **15** write down the size of the angle marked with a letter:

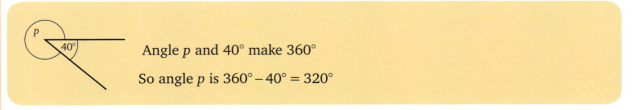

p

40°

Angle *p* and 40° make 360°

So angle *p* is 360° − 40° = 320°

11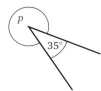

13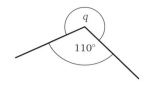

15

12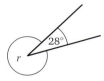

14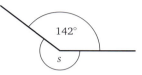

Find the following angles:

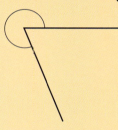

This is a reflex angle and it is bigger than 3 right angles, i.e. it is greater than 270°.

To find this angle, we need to measure the smaller angle, marked *p*.

Angle *p* is 68° so the reflex angle is 360° − 68° = 292°.

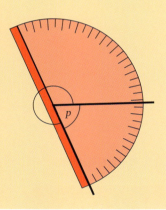

Find:

16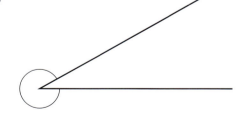

17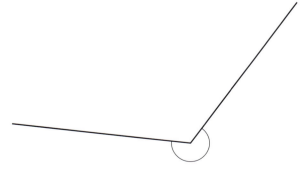

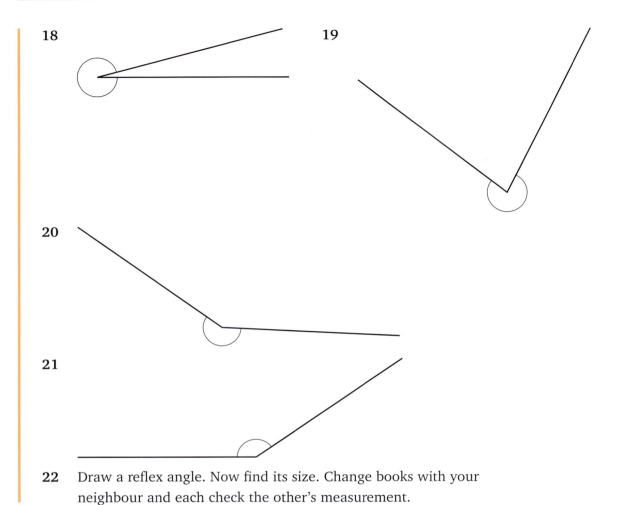

18

19

20

21

22 Draw a reflex angle. Now find its size. Change books with your neighbour and each check the other's measurement.

Mixed questions

Exercise 10h

Use a clock diagram to draw the angle that the *minute* hand of a clock turns through in the following times. In each question write down the size of the angle in degrees.

1	5 minutes	**3**	15 minutes	**5**	25 minutes
2	10 minutes	**4**	20 minutes	**6**	30 minutes

The seconds hand of a clock starts at 12. Which number is it pointing to when it has turned through an angle of:

7	90°	**11**	150°	**15**	420°	**19**	540°
8	60°	**12**	270°	**16**	180°	**20**	240°
9	120°	**13**	30°	**17**	450°	**21**	390°
10	360°	**14**	300°	**18**	210°	**22**	720°

If you start by facing north and turn clockwise, draw a sketch to show roughly the direction in which you are facing if you turn through:

60°

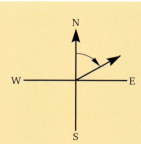

23	45°	27	200°	31	270°
24	70°	28	300°	32	10°
25	120°	29	20°	33	80°
26	50°	30	100°	34	250°

Start by drawing the four compass directions like the diagram above.

Estimate the size, in degrees, of each of the following angles:

35

39

43

36

40

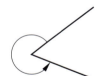

44

37

41

45

38

42

46

Draw the following angles as well as you can by estimating, i.e. without using a protractor. Use a clockface if it helps. Then measure your angles with a protractor.

47	45°	**50**	30°	**53**	150°	**56**	20°	**59**	330°
48	90°	**51**	60°	**54**	200°	**57**	5°	**60**	95°
49	120°	**52**	10°	**55**	290°	**58**	170°	**61**	250°

? Puzzle

This is an exercise for two people. You need a good map and a protractor.

Toss a coin to see who takes the first turn.

One player finds two places on the map. This player shows the other player the position of one place on the map and gives the direction of the second place from the first by estimating the angle that must be turned through clockwise from north. If the second player finds this place within 10 seconds, he has won and it is his turn. Otherwise the first player has won and gets another turn. Play as many times at you wish.

The winner is the player with the most successes.

Any disputes about the direction given are solved by measuring the actual angle with a protractor. Directions within 10° are acceptable. If 10 seconds is not long enough, increase the time allowed to 15 seconds.

Drawing angles using a protractor

To draw an angle of 120° start by drawing one arm and mark the vertex.

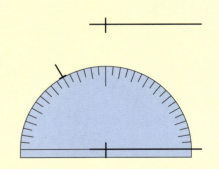

Place your protractor as shown in the diagram.
Make sure that the vertex is at the centre of the base line.

Choose the scale that starts at 0° on your drawn line and mark the paper next to the 120° mark on the scale.

Remove the protractor and join your mark to the vertex.

Now look at your angle: does it look the right size?

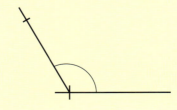

Exercise 10i

Use your protractor to draw the following angles accurately:

1	25°	**4**	160°	**7**	110°	**10**	125°	**13**	105°
2	37°	**5**	83°	**8**	49°	**11**	175°	**14**	136°
3	55°	**6**	15°	**9**	65°	**12**	72°	**15**	90°

Change books with your neighbour and measure each other's angles
as a check on accuracy.

Exercise 10j

1 Draw a diagram showing the two angles that you turn through if you
start by facing north and then turn clockwise through 60°, stop for a
moment and then continue turning until you are facing south. What is
the sum of these two angles?

2 Draw a clock diagram to show the two angles turned through by
the seconds hand if it is started at 2, stopped at 6, started again and
finally stopped at 8. What is the sum of these two angles?

3 Draw an angle of 180°, without using your protractor.

Angles on a straight line

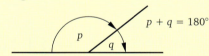

$p + q = 180°$

Angles on a straight line add up to 180°.

Parallel lines

Two lines are parallel when they are always the same distance apart.
Parallel lines are marked with arrows.

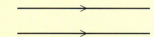

Two lines intersect when they cross on a flat surface.
(A flat surface is called a plane surface.)

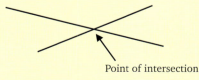

Point of intersection

Perpendicular lines intersect at right angles. A right angle is marked with a square.

Exercise 10k

In questions **1** to **12** calculate the size of the angle marked with a letter:

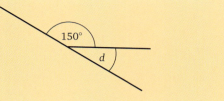

Angles d and 150° together make a straight line

So $d + 150° = 180°$

$\therefore \quad d = 30°$

1

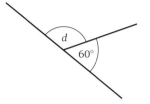

5

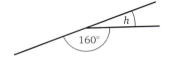

9

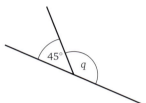

2

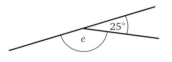

6

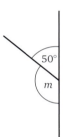

10

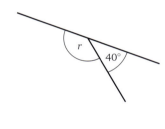

3

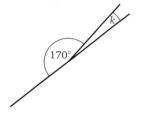

7

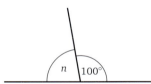

11

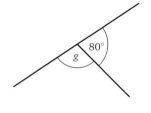

4

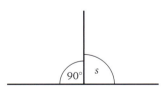

8

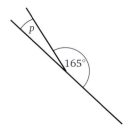

12

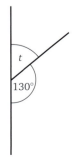

A, B, C and D are points.

Describe the line segments

a AB and AD   **b** AB and DC

c calculate the size of angle *p*.

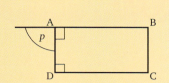

a AB and AD are perpendicular.

b AB and DC are parallel.

c *p* = 90° (angles on a straight line).

13 Name two different pairs of parallel lines.

14 **a** Name a pair of parallel lines.

   b Find the size of angle *s*.

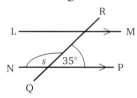

15 **a** Name a pair of perpendicular lines.

   b Find the size of angle *q*.

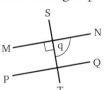

40° and angle *p* make a complete revolution.

In questions **16** to **21** find the sizes of the marked angles.

16

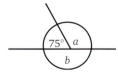

17

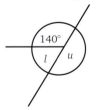

18

19

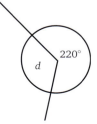

20

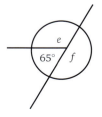

21

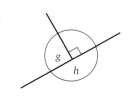

Investigation

1 How do you find the angle turned through
by the hour hand of a clock in a given time?

  a Start by finding the angle turned through in 12 hours,
     then the angle for any other complete number of hours.

  b Next find the angle turned through for any fraction
     of an hour and, lastly, through a number of minutes.

2 Extend your investigation to the minute hand and the
  seconds hand.

3 How do you find the angle between the hands of a clock at any time?
  Start with times that give you the angles that are easiest to find.
  Remember that at 4.30 the minute hand will point to the 6 and the
  hour hand will be exactly half way between 4 and 5.

4 Find out how many times there are in a day when the angle between
  the hour hand and the minute hand has a particular value, say 90°,
  180° or 120°.

5 What happens if the clock loses 10 minutes each hour?
  How many degrees would the minute hand turn through in 1 hour, or
  15 minutes or any other time?

6 What happens if the clock gains 5 minutes every hour?

Mixed exercises

Intersecting lines

Exercise 10I

  1 What angle does the minute hand of a clock turn through when it
     moves from 1 to 9?

  2 Draw an angle of 50°.

  3 Estimate the size of this angle:

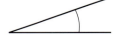

4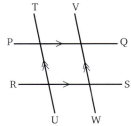

  a Write down one pair of parallel lines.

  b Write down one pair of intersecting lines.

5 Write down the size of the angle marked *s*.

6 Find each of the equal angles marked *e*.

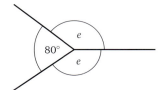

Exercise 10m

1 What angle does the minute hand of a clock turn through when it moves from 10 to 6?

2 If you start facing north and turn clockwise through an angle of 270°, in which direction are you then facing?

3 Measure the angle marked *q*.

4 Write down the sizes of the angles marked *f* and *g*.

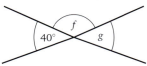

5 Write down the size of the angle marked *h*.

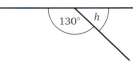

6 Angles *p* and *q* are angles on a straight line. Angle *p* is five times the size of angle *q*. What is the size of angle *q*?

In this chapter you have seen that...

✔ a line goes on forever in both directions

✔ a line that has one end is called a ray and a line with two ends is called a line segment.

✔ a revolution can be divided into 4 right angles

✔ a revolution can be divided into 360°

✔ an acute angle is smaller than 90°

✔ an obtuse angle is larger than 90° but smaller than 180°

✔ a reflex angle is larger than 180°

✔ angles on a straight line add up to 180°

✔ parallel lines are always the same distance apart

✔ perpendicular lines are at right angles to each other.

At the end of this chapter you should be able to...

1 Identify axes of symmetry in given shapes.

2 Identify shapes that have rotational symmetry.

3 Draw axes of symmetry in given shapes.

4 Identify:
   a isosceles triangles   **c** rhombuses
   b equilateral triangles   **d** congruent shapes.

You need to know...

✔ how to use a ruler to draw straight lines

✔ how to use tracing paper

Key words

axis of symmetry, congruent, isosceles, line symmetry, rhombus, rotate, rotational symmetry, symmetry, triangle

To most people symmetrical objects are a pleasure to look at. These shapes are all around us, whether they are man-made things like a jumbo jet or a suspension bridge, or shapes that occur in nature such as a snowflake or an open flower.

Line symmetry

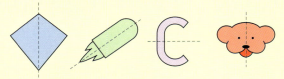

These four shapes are *symmetrical*. If they were folded along the broken line, one half of the drawing would fit exactly over the other half.

Fold a piece of paper in half and cut a shape from the folded edge. When unfolded, the resulting shape is symmetrical. The fold line is the *axis of symmetry*.

Exercise 11a

Some of the shapes below have one axis of symmetry and some have none.
State which of the drawings **1** to **6** have an axis of symmetry.

1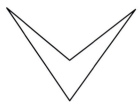

3

5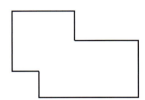

2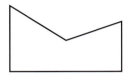

4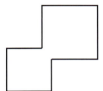

6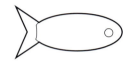

Copy the following drawings on squared paper and complete them so that the
broken line is the axis of symmetry:

7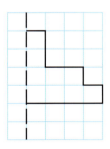

9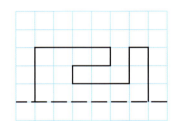

11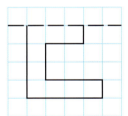

8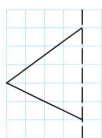

10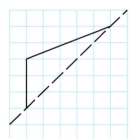

12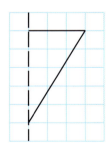

Congruency

This shape has a line of symmetry.

If we cut the shape along the line of symmetry, we get two identical shapes.

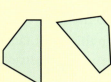

If we turn one over and rotate it, it is still identical to the other shape.

When two shapes are exactly the same shape and size, they are called congruent shapes.

ⓘ Investigation

1 Which of these shapes are congruent?

a

b

c

d

e

f

g

h

i

2 Draw a clock face without hands, like the one below.

Although there are no figures on the clock face assume that the 12 is at the top, that is, in its normal position.

a How many axes of symmetry does this face have?

b How many axes of symmetry would it have if the position of the 12 was marked with a double line?

c Repeat parts **a** and **b** if the figures 1 to 12 are added to the face.

3 The time on this clock face is 6 o'clock.
What we see is symmetrical about a vertical line drawn through the centre of the clock.
Can you find any other time when the position of the hands is symmetrical about this vertical line through the centre?
Would your answer be the same if the hour hand and minute hand were the same length?
Illustrate your answers with sketches.

4 The sketch shows that the time is 2.25.
Make a copy and mark the position of the hour hand and the position of the minute hand when they are reflected in the broken line. (It is probably better to do this in a different colour.) Do the reflected hands give an acceptable time? If so, what time is it?

5

Actual time	'Reflected' time
1.55	
2.50	
3.45	
4.40	
5.35	
6.30	
7.25	
8.20	
9.15	
10.10	
11.05	

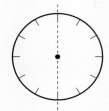

Copy the table and for each line shown:

a draw this clock face and mark in the time

b mark (in a different colour) the position of the hands if they are reflected in the broken line

c complete the corresponding 'reflected' time in the table.

Can you see a connection between the actual times and the 'reflected' times?

6 Copy this clock face and draw a horizontal broken line through the centre.
Mark the time 1.10.
Draw the position of the hands if they are reflected in the broken line.
Is the 'reflected' time an acceptable time?
Justify your answer.

Two axes of symmetry

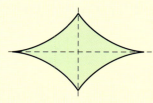 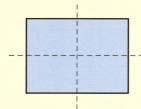 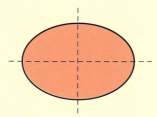

In these shapes there are two lines along which it is possible to fold the paper so that one half fits exactly over the other half.

Fold a piece of paper twice, cut a shape as shown and unfold it. The resulting shape has two axes of symmetry.

Exercise 11b

How many axes of symmetry are there in each of the following shapes?

1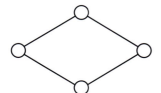

3

5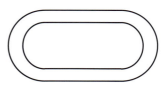

2

4

6

Copy the following drawings on squared paper and complete them so that the two broken lines are the two axes of symmetry.

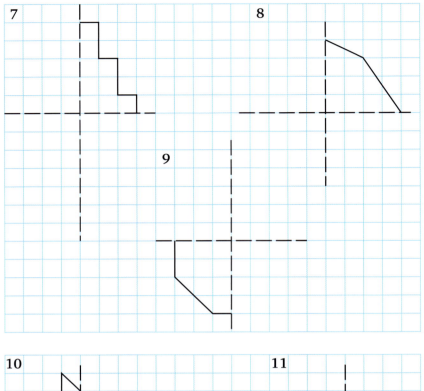

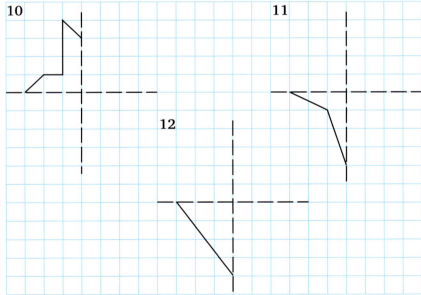

Three or more axes of symmetry

It is possible to have more than two lines of symmetry.

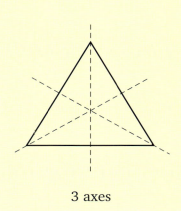
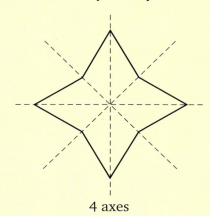
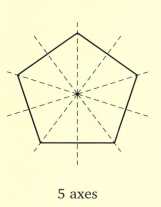

3 axes   4 axes   5 axes

Exercise 11c

How many axes of symmetry are there in each of the following shapes?

1

3

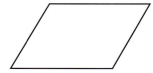

2

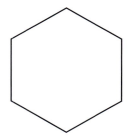

4

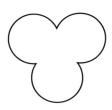

5

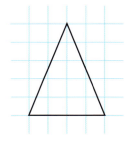

Copy the triangle on squared paper and mark in the axis of symmetry.

A triangle with an axis of symmetry is called an *isosceles triangle*.

6

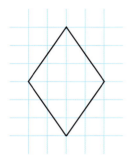

Copy the quadrilateral on squared paper and mark in the two axes of symmetry.

This quadrilateral (which has four equal sides) is called a *rhombus*.

7   Trace the triangle. Draw in its axes of symmetry.

Measure its three sides.

This triangle is called an *equilateral triangle*.

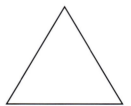

 Puzzle

Fold a square piece of paper twice then fold it a third time along the broken line. Cut a shape, simple or complicated, and unfold the paper. How many axes of symmetry does it have?

   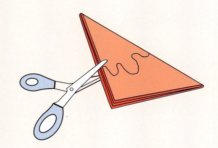

! **Investigation**

Two squares can be put together to give just one shape that has at least one line of symmetry.

There are two ways in which three squares can be arranged to give shapes that have at least one line of symmetry.

Investigate how many different shapes can be made that have at least one line of symmetry, when 4, 5, 6 ... squares are used to make a shape.

Is there any connection between the number of squares used and the number of different shapes that satisfy the given condition?

Rotational symmetry (s-symmetry)

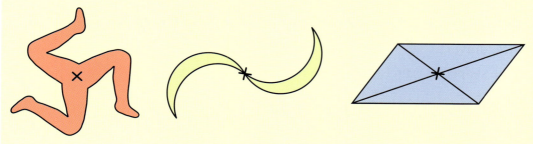

These shapes have a different type of symmetry. They cannot be folded in half but can be turned or rotated about a centre point (marked with ×) and still look the same.

Exercise 11d

1 Lay a piece of tracing paper over any one of the shapes above, trace it and turn it about the cross until it fits over the shape again.

Which of the following shapes have rotational symmetry?

2 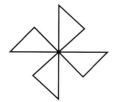   **3** 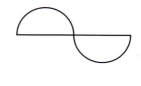   **4**

5    **6** 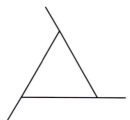   **7**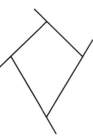

Some shapes have both line symmetry and rotational symmetry:

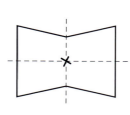   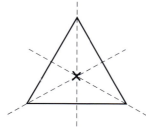   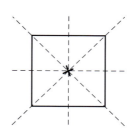

8 Sketch the capital letters of the alphabet. Mark
any axes of symmetry and the centre of rotation if
it exists.

For instance, draw H.

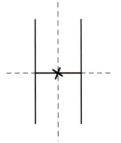

9 Which of the shapes in exercise **11c** have rotational symmetry?

? Puzzle

1 Use the letters I, M, W, S and S to form a five-letter word that has
rotational symmetry – i.e. it reads exactly the same upside down.

2 An important historical building in the West Indies was blown down in
a hurricane.
When it was re-built the contractors placed the original foundation
stone upside down; yet nobody noticed. What was the date of the
original building?

In this chapter you have seen that...

✔ some shapes have line symmetry, some have rotational symmetry, others
have both but some have neither

✔ congruent shapes are exactly the same shape and size

✔ many shapes have more than one axis of symmetry and several of these
also have rotational symmetry.

At the end of this chapter you should be able to...

1 Express given percentages as fractions and vice versa.

2 Express percentages as decimals.

3 Solve problems involving percentages.

4 Express one quantity as a percentage of another.

5 Calculate a percentage of a given quantity.

You need to know...

✔ how to simplify fractions

✔ how to change a mixed number to an improper fraction

✔ how to multiply by a fraction

✔ how to find one quantity as a fraction of another quantity

✔ the meaning of decimals

✔ how to multiply and divide fractions and decimals by 100

✔ the units of length, area and mass

✔ how to change units

Key words

decimal, fraction, percentage

Expressing percentages as fractions

'Per cent' means per hundred, i.e. if 60 per cent of the workers in a factory are women it means that 60 out of every 100 workers are women. If there are 700 workers in the factory, $60 \times 7 = 420$ are women, while if there are 1200 workers, $60 \times 12 = 720$ are women.

In mathematics we are always looking for shorter ways of writing statements and especially for symbols to stand for words. The symbol that means 'percent' is %, i.e. 60 per cent and 60% have exactly the same meaning.

60 per cent means 60 per hundred and this can be written as the fraction $\frac{60}{100}$ $\left(\text{or } \frac{3}{5}\right)$

i.e. 60% of a quantity is exactly the same as $\frac{60}{100}$ $\left(\text{or } \frac{3}{5}\right)$ of that quantity.

If there are 800 cars in a car park and 60% of them are Japanese, then $\frac{60}{100}$ of the cars are Japanese.

i.e.   the number of Japanese cars is $\frac{60}{100} \times 800 = 480$

Exercise 12a

Express   **a** 40%   **b** $22\frac{1}{2}\%$ as fractions in their lowest terms.

a $40\% = \frac{40}{100} = \frac{2}{5}$ **b** $22\frac{1}{2}\% = \frac{45}{2}\% = \frac{45}{2 \times 100} = \frac{9}{40}$

Express as fractions in their lowest terms:

1	20%	**8**	50%	**15**	70%	**22**	95%
2	45%	**9**	65%	**16**	75%	**23**	15%
3	25%	**10**	56%	**17**	48%	**24**	8%
4	72%	**11**	37%	**18**	69%	**25**	82%
5	$33\frac{1}{3}\%$	**12**	$66\frac{2}{3}\%$	**19**	$37\frac{1}{2}\%$	**26**	$87\frac{1}{2}\%$
6	$12\frac{1}{2}\%$	**13**	$62\frac{1}{2}\%$	**20**	$5\frac{1}{3}\%$	**27**	$6\frac{1}{4}\%$
7	$2\frac{1}{2}\%$	**14**	125%	**21**	$17\frac{1}{2}\%$	**28**	150%

Express   **a** 54%   **b** $6\frac{1}{2}\%$ **c** $27\frac{1}{4}\%$ as decimals.

a $54\% = \frac{54}{100} = 0.54$

b $6\frac{1}{2}\% = \frac{6.5}{100} = 0.065$

c $27\frac{1}{4}\% = \frac{109}{4}\% = \frac{109}{4 \times 100} = 0.2725$

Express the following percentages as decimals:

29	47%	**32**	145%	**35**	30%	**38**	$48\frac{1}{2}\%$
30	12%	**33**	$58\frac{3}{4}\%$	**36**	$62\frac{1}{4}\%$	**39**	92%
31	$5\frac{1}{2}\%$	**34**	58%	**37**	350%	**40**	65%

41	120%	**43**	$85\frac{5}{8}\%$	**45**	3%	**47**	$5\frac{1}{5}\%$
42	231%	**44**	8%	**46**	180%	**48**	$54\frac{1}{8}\%$

Expressing fractions as percentages

If $\frac{4}{5}$ of the pupils in a school have been away for a holiday, it means that 80 in every 100 have been on holiday,

i.e. $\frac{4}{5}$ is the same as 80%.

A fraction may be converted into a percentage by multiplying that fraction by 100%. This does not alter its value, since 100% is 1.

A decimal may be converted into a percentage by multiplying it by 100%.

Exercise 12b

Express $\frac{7}{20}$ as a percentage.

$$\frac{7}{20} = \frac{7}{20} \times \overset{5}{100}\% = 35\%$$

Express the following fractions as percentages:

1	$\frac{1}{2}$	**5**	$\frac{21}{40}$	**9**	$\frac{3}{8}$	**13**	$\frac{7}{5}$	**17**	$\frac{7}{20}$
2	$\frac{7}{10}$	**6**	$\frac{1}{4}$	**10**	$\frac{23}{50}$	**14**	$\frac{5}{8}$	**18**	$\frac{31}{25}$
3	$\frac{13}{20}$	**7**	$\frac{3}{20}$	**11**	$\frac{3}{4}$	**15**	$\frac{47}{50}$	**19**	$\frac{7}{8}$
4	$\frac{1}{3}$	**8**	$\frac{4}{25}$	**12**	$\frac{9}{20}$	**16**	$\frac{3}{5}$	**20**	$\frac{8}{5}$

Express **a** 0.7 **b** 1.24 as percentages.

a $0.7 = 0.7 \times 100\% = 70\%$ **b** $1.24 = 1.24 \times 100\% = 124\%$

Express the following decimals as percentages:

21	0.5	**25**	0.625	**29**	2.64	**33**	1.25	**37**	0.16
22	0.22	**26**	0.9	**30**	0.845	**34**	3.41	**38**	1.39
23	0.83	**27**	0.04	**31**	0.25	**35**	0.075	**39**	6.35
24	1.72	**28**	0.55	**32**	0.74	**36**	0.36	**40**	0.1825

Exercise 12c

1 Express as fractions in their lowest terms:
     a   30%   **b**   85%   **c** $42\frac{1}{2}\%$ **d** $5\frac{1}{4}\%$

2 Express as decimals:
     a   44%   **b**   68%   **c**   170%   **d** $16\frac{1}{2}\%$

3 Express as percentages:
     a $\frac{2}{5}$ **b** $\frac{17}{20}$ **c** $\frac{1}{8}$ **d** $\frac{17}{16}$

4 Express as percentages:
     a   0.2   **b**   0.62   **c**   0.845   **d**   1.78

Copy and complete the following table:

	Fraction	Percentage	Decimal
	$\frac{3}{4}$	75%	0.75
5	$\frac{4}{5}$		
6		60%	
7			0.7
8	$\frac{11}{20}$		
9		44%	
10			0.32

Problems

Suppose that in the town of Doxton 25 families in every 100 own a car. We can deduce from this that 75 in every 100 families do not. Since every family either owns a car or does not own a car, if we are given one percentage we can deduce the other.

Exercise 12d

If 56% of homes have a telephone, what percentage do not?

All homes (i.e. 100% of homes) either have, or do not have, a telephone.

If 56% have a telephone, then (100 − 56)% do not,

i.e. 44% do not.

1   If 48% of the pupils in a school are girls, what percentage are boys?

2   If 87% of households have a television set, what percentage do not?

3   In the fourth year, 64% of the pupils do not study chemistry. What percentage study chemistry?

4   In a box of oranges, 8% are bad. What percentage are good?

5   Twelve per cent of the persons taking a driver's test fail to pass first time. What percentage pass first time?

6   A hockey team won 62% of their matches and drew 26% of them. What percentage did they lose?

7   A soccer team drew 12% of their matches and lost 45% of them. What percentage did they win?

8   Deductions from a youth's wage were: income tax 18%, other deductions 14%. What percentage did he keep?

9   In an election 40% of the electorate voted for Mrs Long, 32% for Mr Singh and the remainder voted for Miss Berry. What percentage voted for Miss Berry if there were only three candidates and 8% of the electorate failed to vote?

10 In a school 36% of the pupils study French and 38% study German. If 12% study both languages, what percentage do not study either?

11 Eighty-five per cent of the first year pupils in a school study craft and 72% study photography. If 60% study both subjects, what percentage study neither?

12 A concert is attended by 1200 people. If 42% are adult females and 37% are adult males, how many children attended?

13 The attendance at an athletics meeting is 14 000. If 68% are men and boys and 22% are women, how many are girls?

14 In a book, 98% of the pages contain text, diagrams or both. If 88% of the pages contain text and 32% contain diagrams, what percentage contain
    a   neither text nor diagrams
    b   only diagrams
    c   only text
    d   both text and diagrams?

Puzzle

Alice in Wonderland was created by Lewis Carroll whose real name was Charles Lutwidge Dodgson. Dodgson was a mathematician who thought up many puzzles. This is one, with the title of 'Casualties':

If 70% have lost an eye, 75% an ear, 80% an arm, 85% a leg, what percentage, at least, must have lost all four?

Expressing one quantity as a percentage of another

If we wish to find 4 as a percentage of 20, we know that 4 is $\frac{4}{20}$ of 20

and $\qquad \frac{4}{20} = \frac{4}{20} \times 100\%$

i.e. 4 as a percentage of 20 is

$$\frac{4}{20} \times 100\% = 20\%$$

To express one quantity as a percentage of another, we divide the first quantity by the second and multiply this fraction by 100%.

Exercise 12e

Express 20 cm as a percentage of 3 m.

(First express 3 m in centimetres to bring both quantities to the same unit.)

$$3\,\text{m} = 3 \times 100\,\text{cm} = 300\,\text{cm}$$

Then the first quantity as a percentage of the second quantity is

$$\frac{20}{300} \times 100\% = \frac{20}{3}\% = 6\frac{2}{3}\%$$

Express the first quantity as a percentage of the second:

1	3, 12	**7**	60 cm, 4 m		
2	30 cm, 50 cm	**8**	10 ft, 40 ft		
3	3 m, 9 m	**9**	5, 50		
4	4 in, 12 in	**10**	2 cm, 10 cm	**13**	40, 20
5	15, 20	**11**	600 m, 2 km	**14**	35 m, 56 m
6	24 cm, 40 cm	**12**	$3\frac{1}{2}$ yd, 7 yd	**15**	50 cm, 5 m

Make sure that both quantities are measured in the same unit.

16	8 in, 12 in	**23**	198 mm², 275 mm²	**30**	74 c, $1.11
17	20 m², 80 m²	**24**	50 m², 15 m²	**31**	37 mm, 148 cm
18	75 cm², 200 cm²	**25**	3.6 t, 5 t	**32**	900 g, 2.5 kg
19	25 cm², 125 cm²	**26**	33.6 g, 80 g	**33**	45 c, $1.35
20	4 litres, 10 litres	**27**	1200 g, 3 kg	**34**	98 mm, 2.45 m
21	3 pints, 5 pints	**28**	3.64 kg, 5.6 kg	**35**	4 mm, 3 cm
22	200 mm², 800 mm²	**29**	28 cm, 1.2 m	**36**	84 g, 3.36 kg

Finding a percentage of a quantity

To find a percentage of a quantity, change the percentage to a fraction and multiply by the quantity.

Exercise 12f

Find the value of   **a** 12% of 450   **b** $7\frac{1}{3}$% of 3.75 m

a 12% of 450 $= \frac{12}{100} \times 450 = 54$

b $7\frac{1}{3}$% of 3.75 m $= 7\frac{1}{3}$% of 375 cm $= \frac{22}{3}$% of 375 cm

$\qquad = \frac{22}{3 \times 100} \times 375$ cm $= 27.5$ cm

Find the value of:

1	40% of 120	**13**	45% of 740	**25**	$66\frac{2}{3}$% of 480 m²
2	12% of 800 g	**14**	33% of 600 kg	**26**	$32\frac{1}{7}$% of 140 km
3	74% of 75 cm	**15**	6% of 24 m	**27**	$62\frac{1}{2}$% of 8 km
4	44% of 650 km	**16**	15% of $10	**28**	$74\frac{1}{2}$% of 200 cm²
5	8% of $2	**17**	17% of 2 km	**29**	$33\frac{1}{3}$% of 42 c
6	77% of 4 kg	**18**	32% of 5 litres	**30**	$82\frac{1}{5}$% of $65
7	70% of 360	**19**	30% of $250	**31**	12% of $4
8	86% of 1150 g	**20**	66% of 300 m	**32**	$7\frac{1}{2}$% of 80 g
9	55% of 8.6 m	**21**	$33\frac{1}{3}$% of 270 g	**33**	$2\frac{1}{3}$% of 90 m
10	96% of 215 cm²	**22**	$5\frac{1}{4}$% of 56 mm	**34**	$16\frac{2}{3}$% of $60
11	63% of 4 m	**23**	$37\frac{1}{2}$% of 48 cm	**35**	$3\frac{1}{8}$% of 64 kg
12	96% of 15 m²	**24**	$22\frac{1}{2}$% of 40 m²	**36**	$87\frac{1}{2}$% of 16 mm

Investigation

Banks offer many different accounts. Get an up-to-date leaflet from one bank that gives details of all its different accounts and the rate of interest offered on each.

1 Write a short report on which account you would use, and why, if
   a you are saving to buy a pair of trainers
   b a relative has given you $10 000 and you want to keep it safe until you leave school.

Problems

Exercise 12g

In the second year, 287 of the 350 pupils study geography. What percentage study geography?

Express 287 as a fraction of 350, then multiply by 100%.

$$\text{Percentage studying geography} = \frac{287}{350} \times 100\%$$

$$= 82\%$$

1 There are 60 boys in the third year, 24 of whom study chemistry. What percentage of third year boys study chemistry?

 Read the question carefully to make sure that you understand what you are being asked to find. Read it several times if necessary.

2 In a history test, Pauline scored 28 out of a possible 40. What was her percentage mark?

3 Out of 20 drivers tested in one day for a driver's licence, 4 of them failed. What percentage failed?

4 There are 60 photographs in a book, 12 of which are coloured. What is the percentage of coloured photographs?

5 Forty-two of the 60 choristers in a choir wear spectacles. What percentage do not?

6 Each week a boy saves $3000 of the $12 000 he earns. What percentage does he spend?

7 A secretary takes 56 letters to the post office for posting; 14 are registered and the remainder are ordinary mail. What percentage go by ordinary mail?

8   Judy obtained 80 marks out of a possible 120 in her end of term maths examination. What was her percentage mark?

9   Jane's gross wage is $12 000 per week, but her 'take home' pay is only $7800. What percentage is the take home pay of her gross wage?

10   If 8% of a crowd of 24 500 at a football match were females, how many females attended?

If 54% of the 1800 pupils in a school are boys, how many girls are there in the school?

Method 1   First find the number of boys in the school: this is 54% of 1800.

$$\text{Number of boys} = \frac{54}{100} \times 1800$$
$$= 972$$

Now you can find the number of girls:

$$\text{Number of girls} = 1800 - 972$$
$$= 828$$

Method 2   As 54% of the pupils are boys, 100% − 54%, i.e. 46% of the pupils are girls.

So the number of girls is 46% of 1800: $\frac{46}{100} \times 1800 = 828$

11   In a garage, 16 of the 30 cars which are for sale are second hand. What percentage of the cars are
    a   new   b   second hand?

12   There are 80 houses on Freedom Street and 65% of them have a telephone. How many houses
    a   have a telephone   b   do not have a telephone?

13   In my class there are 30 pupils and 40% of them have a bicycle. How many pupils
    a   have a bicycle   b   do not have a bicycle?

14   Yesterday, of the 150 flights leaving London Airport, 2% were bound for Kingston. How many of these flights
    a   flew to Kingston   b   did not fly to Kingston?

15   In a particular year, 64% of the 16 000 Jewish immigrants into Israel came from Eastern Europe. How many of the immigrants did not come from Eastern Europe?

16 There are 120 shops in the High Street, 35% of which sell food. How many High Street shops do not sell food?

17 Last year the amount I paid in Land Tax was $520. This year my rates will increase by 12%. Find the increase.

18 A mathematics book has 320 pages, 40% of which are on algebra, 25% on geometry and the remainder on arithmetic. How many pages of arithmetic are there?

Mixed exercises

Exercise 12h

1 Express as a fraction in its lowest terms:
   a 40%   b 54%   c $27\frac{1}{2}\%$

2 Express as a percentage:
   a $\frac{3}{5}$ b 0.78   c 0.125

3 Express 2 m as a percentage of 25 m.

4 Express 25 c as a percentage of $2.

5 Find 45% of 120 millilitres.

6 If 3% of telephone calls are connected to the wrong number, what percentage of calls are connected to the correct number?

Exercise 12i

1 Express 36%
   a as a vulgar fraction in its lowest terms
   b as a decimal.

2 Express as a percentage
   a $\frac{5}{8}$ b $1\frac{1}{4}$ c 2.5

3 Express 250 g as a percentage of 2 kg.

4 Find 85% of 340 m².

5 The cost of insuring a car is about 8% of its value. Find the cost of insuring a car valued at $550 000.

Exercise 12j

1 Find the first quantity as a percentage of the second quantity
   a 10 m, 80 m   **b** 75 c, $2   **c** 150 cm, 3 m

2 Express as a percentage
   a $\frac{2}{5}$ **b** 0.279   **c** $1\frac{5}{8}$

3 Express $12\frac{1}{2}\%$ as

   a a vulgar fraction in its lowest terms   **b** a decimal.

4 Find 36% of $2.50.

5 There are 450 children in a primary school, 12% of whom do not speak English at home. Find the number of children for whom English is not their home language.

 Puzzle

If 5th September falls on a Friday, on which day of the week will Christmas Day fall?

In this chapter you have seen that...

✔ a percentage can be expressed as a fraction by putting it over 100 and simplifying, e.g. $70\% = \frac{70}{100} = \frac{7}{10}$

✔ a percentage can be expressed as a decimal by dividing it by 100, e.g. $65\% = 65 \div 100 = 0.65$

✔ a fraction, or a decimal, can be expressed as a percentage by multiplying it by 100, e.g. $\frac{2}{5} = \frac{2}{5} \times 100\% = 40\%$ and $0.8 = 0.8 \times 100\% = 80\%$

✔ to express one quantity as a percentage of another, first express the quantity as a fraction of the second and multiply by 100, e.g. 4 as a percentage of 20 is $\frac{4}{20} \times 100\% = 20\%$

✔ to find a percentage of a quantity, multiply the percentage by the quantity and divide by 100, e.g. 20% of 80 cm $= \frac{20}{100} \times 80 \text{ cm} = 16 \text{ cm}$

13 Money matters

At the end of this chapter you should be able to...

1 Solve problems using units of money from different countries.
2 Find the cost of a number of articles given the cost of one.
3 Find the unit cost of an article given the cost of many.
4 Work out best buys.
5 Work out the profit/loss on buying and selling goods.
6 Find the percentage gain/loss when things are bought and sold.

You need to know how to...

✔ multiply and divide by whole numbers
✔ multiply and divide using decimals
✔ express one quantity as a percentage of another

Key words

best buys, loss, percentage, profit

Money

Money is the essence of everyday life. We use money to buy what we need – food, shelter, entertainment – and we earn money by getting paid for our labour or by selling things we own. Money enables us to exchange our labour or goods for services or other goods.

We can have money in notes and coins but our wealth may also be shown in numbers on paper or in a computer. Since almost everything has value, all we have – our home, its contents, our bicycle or our football – can be translated into a sum of money. The sum total gives a measure of our wealth and we can exchange some of it for goods and services at any time.

Money units

Many countries use units of money that are divided into hundredths. For example

Eastern Caribbean	1 dollar ($) = 100 cents (c)
UK	1 pound (£) = 100 pence (p)
USA	1 dollar ($) = 100 cents (c)
Euro zone	1 euro (€) = 100 cents (c)

Exercise 13a

Express each quantity in terms of the unit given in brackets.

1	7 dollars	(cents)	6	43 dollars 81 cents	(cents)
2	£6	(pence)	7	€11 3c	(cents)
3	€8	(cents)	8	£6 15p	(pence)
4	€13	(cents)	9	£2 10p	(pence)
5	€7 35c	(cents)	10	£5 4p	(pence)

Change 420 c to $

Divide by 100:   420 c = $4.20

(Note that we always give dollars to 2 decimal places so we write $4.20 rather than $4.2. Other currencies are written in the same way.)

11	126 p	(£)	16	228 p	(£)
12	350 cents	(dollars)	17	€3 47c	(euros)
13	190 p	(£)	18	580 c	($)
14	350 cents	(euros)	19	€11 9c	(cents)
15	43 dollars 7 cents	(dollars)	20	£6 8p	(£)

21   In the UK, one tin of baked beans costs 32 p. Find the cost, in pounds, of ten of these tins.

22   Find the total cost, in dollars, of a book costing US$4, a pencil costing 30 cents and a magazine costing 75 cents.

23   Find the cost, in dollars, of 20 litres of petrol at $96 a litre.

24   One can of cola costs 50 cents. Find the cost, in euros, of twelve such cans.

25   Find the cost of ten stamps at 29 c each. If you paid for these stamps with a US$5 note, how much change would you get?

We often want to know how much a large quantity of something will cost when we know the cost of one. At other times we can see the price of a pack but want to know how much one costs.

Exercise 13b

A cricket ball costs $4350. Work out the cost of a box of 12 cricket balls.

1 cricket ball costs $4350
So 12 cricket balls cost $4350 × 12 = $52 200

1   Find the cost of 12 oranges at $9.50 each.

2   The cost of 1 kg of sugar is $150. What is the cost of 6 kg?

3   A carpet costs $980 per square metre. How much will 7.8 square metres of this carpet cost?

4   Work out the cost of 6 pens if they cost $145 each.

5   Lil works 32 hours a week and is paid $670 an hour. What is her basic weekly wage?

6   Wynford's mother told him that his lunchtime treat had cost her $70.50. How much did his mother pay for a pack of 24 treats?

A tray of 36 plants costs $1620. How much is each plant?

36 plants cost $1620
So 1 plant costs $1620 ÷ 36 = $45

7   A pack of 24 chocolate biscuits cost $108. How much did one cost?

8   Mrs Rayburn paid $1710 for 20 litres of gas. What was the price per litre?

9   The cost of running a refrigerator for 3.5 hours is $17.50.
    What is the cost of running this refrigerator for 1 hour?

10 James is paid $25 920 for a basic week of 36 hours. Find his hourly rate.

11 To make a fruit salad for a party Mrs Walcott paid the following amounts for fruit:
    oranges $280, mangos $175, cherries $240, pine $120 and pears $337.
    Her fruit salad served 12 people. How much was this per person?

12 Mr Riley spent $2250 on gas when it cost $90 a litre.
    How many litres did he buy?

13   Brian sold 74 m² of turf to Mr Barnes for $50 320.
      How much was this per square metre?

14   Anita's annual salary for 52 weeks is $1 300 000. How much is this
      per week?

15   Blossom is paid $20 330 for a 38-hour week. Work out her hourly
      rate.

<u>16</u>   To produce 1200 booklets a publisher spent $15 400 on materials,
      $27 200 on labour and $15 000 to cover distribution costs. How
      much did it cost to publish each booklet?

<u>17</u>   To run a gymnastics club for an evening it costs $4500 to hire the
      gym and $3900 to pay the teacher. If 14 enthusiasts attend how
      much should each be charged to cover the cost?

<u>18</u>   Lisa wants to paint the inside of her house which has 6 rooms. She
      buys 16 litres of paint at $1600 a litre, 3 brushes at $440 each and
      spends $698 on sand paper and filler.
      a   How much does she spend altogether?
      b   How much does this work out per room?

Best buys

These two jars of coffee contain different amounts of coffee
and cost different amounts of money.

You can find which of these two jars is the better value for
money (or best buy). There are two ways you can do this.

The first way is to work out the cost of the *same mass* for
each jar.

The smaller jar holds 200 g and costs $275.

The larger jar holds 1 kg and costs $1345.

To compare the cost of coffee with the smaller jar, you can find the cost of
200 g.

1 kg = 5 × 200 g, so 200 g of coffee in the larger jar costs $1345 ÷ 5 = $269.

Now you can see that the cost of 200 g of coffee is less in the larger jar than
in the smaller jar. So the larger jar is better value for money.

The second way is to find the mass of coffee that the *same amount of money*
will buy.

The smaller jar costs \$275 for 200 g, so \$1 will buy 200 g ÷ 275 = 0.73 g to the nearest gram. The larger jar costs \$1345 for 1 kg, so \$1 will buy 1 kg ÷ 1345 = 1000 g ÷ 1345 = 0.74 g to the nearest gram.

Now you can see that \$1 will buy more coffee in the larger jar than in the smaller jar. So the larger jar is better value for money.

Exercise 13c

1

3 for \$111   \$29 each

Peppers are sold in packs of three for \$111 or individually for \$29.
Which is the best value?
Give a reason for your answer.

You can compare these prices by either finding the cost of 3 of the single peppers or by finding the cost of one of the peppers in the pack. Whichever way you choose, remember to show your working.

2   Cans of cola are sold in packs of 4 cans for \$240 and in packs of 6 cans for \$348.

Which pack is better value for money?
Give a reason for your answer.

3   These are two different bags of paper clips – 100 clips for \$125 and 75 clips for \$105.
Which bag is better value for money?
Explain your answer.

You can compare the number of clips per cent or the cost of one clip.

100 clips \$125   75 clips \$105

4   There are two different packs of tomatoes – 540 g for \$254 and 1 kg for \$370. Which pack of tomatoes is better value for money?
Give a reason for your answer.

540 g \$254   1 kg \$370

5 String is sold in rolls of two sizes – 500 m
   for $130 and 2 km for $500.
   Jane said that the larger roll is better
   value for money.
   Is Jane correct?
   Give a reason for your answer.

6 Cheese is sold from the delicatessen counter where
   it is priced at $352 per 500 g.

   Cheese is also sold in prepacks weighing 200 g and
   costing $130 a pack.

   Which way of buying cheese is better value?

   Explain your answer.

7 Coffee is sold in jars of 75 g for $120 and 200 g for $290.
   a   What is the cost of 25 g from the smaller jar?
   b   What is the cost of 25 g from the larger jar?
   c   Which jar is the better value for money?
   d   An even larger jar contains 250 g grams of coffee
       and costs $370.

       Is this better value for money than either of the other two?
       Give a reason for your answer.

8 Which jar is better value?

Profit and loss

If you buy an article for $100 and sell it for $120 you have made a gain or
profit of $20.
However, if you sell the same article for $75 you have made a loss of $25.

Remember   selling price – cost price = profit or loss
For short we write Profit = SP – CP
                and Loss = CP – SP

Exercise 13d

1 A calculator is bought for $1200 and sold for $1750. Find the profit.

2 A bookseller buys a book for $1700 and sells it for $3195. Find his profit.

3 A house bought for $7 200 000 is sold for $6 600 000. Find the loss.

4 Find the profit or loss given that
   a the cost price is $340 and the selling price is $476
   b the cost price is $76 500 and the selling price is $103 275
   c the cost price is $440 and the selling price is $330
   d the cost price is $1540 and the selling price is $132.

5 Imported bicycles cost a dealer $24 500 and are sold at $36 000.
   Work out the profit.

6 Six months after paying $800 000 for a car the owner is forced to sell
   for $656 000. How much did he lose?

7 At a cricket club dinner 62 members sat down to eat. The costs
   for the club were: first course $44 640, main course $59 520, dessert
   $42 800, drinks $31 600.
   a Work out the cost of the dinner per head.
   b Members were charged $2995 a head. Did the club make a profit?
      Justify your answer.

8 A supermarket bought 100 pastry cases at $54 each and priced them for
   sale at $89 each. Unfortunately 15 got damaged and were unfit for sale.
   Find
   a the cost of the pastry cases
   b the total income from the sale
   c the total profit
   d what the profit would have been if they could have sold all the
      cases.

Percentage profit and loss

When a retailer buys goods and is able to sell them at a higher price, a
profit is made. This is the difference between the selling price (SP) and the
cost price (CP). The percentage profit or loss made is always calculated by
expressing the profit or loss *as a percentage of the cost price*.

Percentage profit = $\dfrac{\text{profit}}{\text{cost price}} \times 100$ i.e. % profit = $\dfrac{\text{SP} - \text{CP}}{\text{CP}} \times 100$

Similarly % loss = $\dfrac{\text{loss}}{\text{CP}} \times 100$ i.e. % loss = $\dfrac{\text{CP} - \text{SP}}{\text{CP}} \times 100$

Exercise 13e

A shopkeeper bought a clock for $5500 and sold it for $7700. Find his percentage profit.

$$\text{Profit} = \text{SP} - \text{CP}$$
$$= \$7700 - \$5500$$
$$= \$2200$$

$$\% \text{ profit} = \frac{\text{profit}}{\text{CP}} \times 100$$
$$= \frac{\$2200}{\$5500} \times 100 = \frac{2}{5} \times 100$$

i.e. Percentage profit is 40%

1   A box of paints bought for $1200 is sold for $1740. Find the percentage profit.

2   A silversmith sells a silver brooch for $26400. If the brooch cost him $12000 find his percentage profit.

3   An art dealer bought a picture for $350000 and sold it for $630000. Work out his percentage profit.

4   Record albums bought for $1400 are sold at $980. Work out the percentage loss.

5   Ray George sold some vegetables for $189 that he had paid $225 for. Work out his percentage loss.

6   A discount store bought a suite of furniture for $252000 and sold it for $308700. Find the percentage profit.

7   Ernie Pugh buys a dog for $36000 and sells it for $31680. Calculate his percentage loss.

8   A greengrocer buys a box of 150 oranges for $4050 and sells them at $40.50 each.
    Find **a** his profit **b** his percentage profit.

9   A scrap metal dealer buys lead for $7200 and sells it for $13320. Find his percentage profit.

10 By selling a picture for $216000 an antique dealer makes a profit of $172800.
    Find **a** the cost price **b** her percentage profit.

11 Constance bought a necklace for $750 and sold it at a loss of $270.
    a How much did she sell it for? **b** Work out her percentage loss.

12 A second-hand car dealer buys a car for $460000 and sells it at a loss of $69000.
    a How much does he sell it for? **b** Calculate his percentage loss.

13   Dried fruit bought at $12 000 per 50 kg bag is sold at $330 per kilogram.
      Find the percentage profit.

14   Eggs are bought at the farm for $300 per tray and sold at $180 per dozen. If a tray holds 36 eggs, find the percentage profit.

15   A retailer buys 100 articles for $18 000 and sells them at $240 each. Find the percentage profit.

16   A shopkeeper buys 300 articles for $120 000 and sells them at $350 each. Find his percentage loss.

17   The owner of a general store sells an article for $3968, thereby making a profit of $1408.
      Find   **a**   the cost price of the article   **b**   his percentage profit.

18   When a second-hand furniture dealer sold a table for $5280 he made a loss of $720.
      Work out   **a**   the price he paid   **b**   his percentage loss.

19   Elaine buys 35 young trees for $7000 and sells them for $330 each.
      Find   **a**   Elaine's profit   **b**   her percentage profit.

20   Matthew buys a box of 50 old CDs for $6000 and sells them at $200 each.
      Find   **a**   his profit   **b**   his percentage profit.

21   Barrie bought a box of 36 DVDs for $50 000 and sold them at $1250 each.
      a   Did Barrie make a profit or a loss? Justify your answer.
      b   Express your answer as either as percentage gain or a percentage loss.

22   Maria bought a set of 6 dining chairs for $30 000. One of them was damaged but she managed to sell the others at $6750 each. Find her gain or loss percent.

23   Screws are produced at 92 c each and sold in packets of 25 for $36.80. Find the percentage profit.

Mixed exercises

Exercise 13f

1   Work out the cost of 12 footballs at $4350 each.

2   If 1 dollar ($) = 100 cents (c), find
     a $15 in cents   **b**   46 300 cents in dollars.

3   Find the cost of 72 boxes of lego at $695 each.

4   Crunchy peanut butter in Shop A costs $325 for 250 g. In Shop B it costs $408 for 340 g. Which shop gives the better value for money? Explain your answer.

5   A DVD cost a retailer $1320. She sells it for $2046. Find her percentage profit.

6   Pete buys a poster for $360 and sells it to Simon for $288. Find his percentage loss.

Exercise 13g

1   Elsie works a 38-hour week and gets paid $655 an hour. Calculate her weekly pay.

2   Given that 1 pound (£) = 100 pence (p) express
    a   £15.50 in pence   b   650 p in pounds.

3   A tray of 24 plants costs $4560. Work out the cost of one plant.

4   In Mates supermarket a 340 g jar of blackcurrant jam costs $255. In Bestway supermarket a similar jam costs $259 for a 370 g jar. Which jar is the better buy? Explain your answer.

5   A retailer bought 144 tins of baked beans for $7200. He sold them in packs of 4 for $280 a pack. Work out
    a   his total profit   b   his percentage profit.

6   By selling a tea set for $5760 a retailer makes a profit of $2160.
    Find   a   the cost price   b   the percentage profit.

In this chapter you have seen that...

✔ the cost of many articles is the cost of one article multiplied by the number of articles

✔ the cost of one article is given by dividing cost of many by the number of articles

✔ you can determine best buys by comparing the cost of one unit

✔ a profit or a loss is the difference between the cost price and the selling price

✔ the percentage profit or loss is the profit or loss as a percentage of the cost price.

14 Coordinates

Plotting points using positive coordinates

There are many occasions when you need to describe the position of an object. For example, telling a friend how to find your house, finding a square in the game of battleships or describing the position of an aeroplane showing up on a radar screen. In mathematics we need a quick way to describe the position of a point.

We do this by using squared paper and marking a point O at the corner of one square. We then draw a line through O across the page. This line is called Ox. Next we draw a line through O up the page. This line is called Oy. Starting from O we then mark numbered scales on each line.

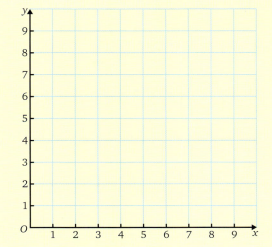

> O is called the origin
> Ox is called the x-axis
> Oy is called the y-axis

We can now describe the position of a point A as follows:

> start from O and move 3 units along Ox,
> then move 5 units up from Ox.

We always use the same method to describe the position of a point:

> start from O, *first* move *along* and *then up*.

We can now shorten the description of the position of the point A to the number pair (3, 5).

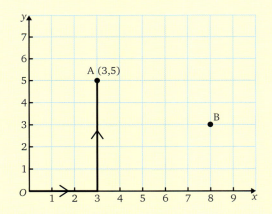

The number pair (3, 5) is referred to as the coordinates of A.

The first number, 3, is called the x-coordinate of A.

The second number, 5, is called the y-coordinate of A.

Now consider another point B

> whose x-coordinate is 8

and > whose y-coordinate is 3.

If we simply refer to the point B (8, 3)

this tells us all that we need to know about the position of B.

The origin is the point (0, 0).

1 Write down the coordinates of the points A, B, C, D, E, F, G and H.

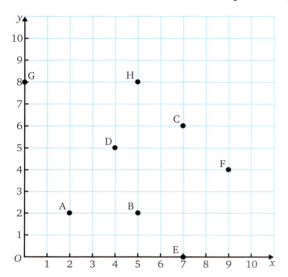

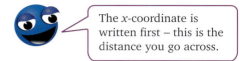

The *x*-coordinate is written first – this is the distance you go across.

2 Draw a set of axes of your own. Along each axis mark points 0, 1, 2, …, 10 units from *O*. Mark the following points and label each point with its own letter:

A(2, 8) B(4, 9) C(7, 9) D(8, 7) E(8, 6) F(9, 4) G(8, 4)
H(7, 3) I(5, 3) J(7, 2) K(7, 1) L(4, 2) M(2, 0) N(0, 2)

Now join your points together in alphabetical order and join A to N.

3 Draw a set of axes and give them scales from 0 to 10. Mark the following points:

A(2, 5) B(7, 5) C(7, 4) D(8, 4) E(8, 3)

F(9, 3) G(9, 2) H(6, 3) I(6, 1) J(7, 1)

K(7, 0) L(5, 0) M(5, 2) N(4, 2) P(4, 0)

Q(2, 0) R(2, 1) S(3, 1) T(3, 2) U(0, 2)

V(0, 3) W(1, 3) X(1, 4) Y(2, 4)

Remember, the first number is the distance you go across and the second number is the distance you go up.

Now join your points together in alphabetical order and join A to Y.

4 Mark the following points on your own set of axes:

A(2, 7) B(8, 7) C(8, 1) D(2, 1)

Join A to B, B to C, C to D and D to A. What is the name of the figure ABCD?

5 Mark the following points on your own set of axes:

A(2, 2) B(8, 2) C(5, 5)

Join A to B, B to C and C to A. Describe fully the triangle ABC.

6   Mark the following points on your own set of axes:

A(4, 0)   B(6, 0)   C(6, 4)   D(4, 4)

Join A to B, B to C, C to D and D to A. What is the name of the figure ABCD?

7   Mark the following points on your own set of axes:

A(5, 2)   B(8, 5)   C(5, 8)   D(2, 5)

Join the points to make the figure ABCD. What is ABCD?

8   On your own set of axes mark the points A(8, 4), B(8, 8) and C(14, 6). Join A to B, B to C and C to A.

Describe fully the figure ABC.

Questions **9** to **14** refer to the points A(1, 7), B(5, 0) and C(0, 14).

9   Write down the x-coordinate of the point B.

10   Write down the y-coordinate of the point A.

11   Write down the x-coordinate of the point C.

12   Write down the x-coordinate of the point A.

13   Write down the y-coordinate of the point C.

14   Write down the y-coordinate of the point B.

Questions **15** to **20** refer to points in the following diagram:

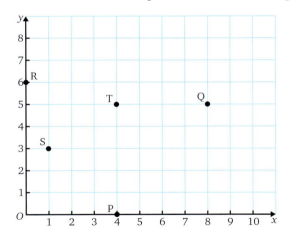

15   Write down the y-coordinate of the point T.

16   Write down the x-coordinate of the point P.

17   Write down the x-coordinate of the point S.

18   Write down the y-coordinate of the point R.

19   Write down the y-coordinate of the point Q.

20   Write down the x-coordinate of the point R.

Questions **21** to **24** refer to the following diagrams:

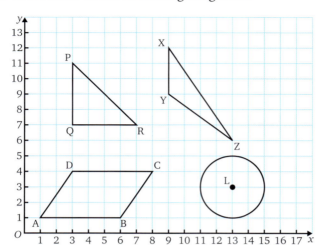

21   Write down the coordinates of the vertices X, Y and Z of triangle XYZ.

22   Write down the coordinates of the vertices of the isosceles triangle PQR. Write down the lengths of the two equal sides.

23   Write down the coordinates of the vertices of the parallelogram ABCD. How long is AB? How long is DC?

24   Write down the coordinates of the centre, L, of the circle. What is the diameter (distance across) of this circle?

For each of the following questions you will need to draw your own set of axes:

25   The points A(2, 1), B(6, 1) and C(6, 5) are three corners of a square ABCD. Mark the points A, B and C. Find the point D and write down the coordinates of D.

26   The points A(2, 1), B(2, 3) and C(7, 3) are three vertices of a rectangle ABCD. Mark the points and find the point D. Write down the coordinates of D.

27   The points A(1, 4), B(4, 7) and C(7, 4) are three vertices of a square ABCD. Mark the points A, B and C and find D. Write down the coordinates of D.

28   Mark the points A(2, 4) and B(8, 4). Join A to B and find the point C, which is the midpoint (the exact middle) of the line AB. Write down the coordinates of C.

29   Mark the points P(3, 5) and Q(3, 9). Join P and Q and mark the point R, which is the midpoint of PQ. Write down the coordinates of R.

30   Mark the points A(0, 5) and B(4, 1). Find the coordinates of the midpoint of AB.

Puzzle

Draw a simple pattern of your own on squared paper but do not show it to anyone. Write down the coordinates of each point and give this set of coordinates to your partner. See if your partner can now draw your diagram.

 Investigation

1   At Newtown Grammar School they always hold important examinations in the school hall. The hall is rectangular and has sufficient space for 8 rows with 16 desks in each row. It is important that each pupil sits in a particular desk.

   a   How can the desks be described so that a particular student can be directed to a particular desk?

   Your system could be all numbers, such as (1, 1), (1, 2), … or all letters – for example, Aa, Ab, … or a mixture of the two – for example, A1, A2, …

   b   Which system is the best for pupils to understand easily and quickly where their place is?

   Explain your answer.

2   What system is used when large numbers of people are to be seated – for example, in a sports stadium or concert hall?

3   How can you remember where your car is parked in an airport carpark where there may be hundreds of cars?

Quadrilaterals

A quadrilateral is bounded by four straight sides. No two of the sides need to be equal and no two of the sides need to be parallel.

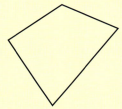

There are, however, some special quadrilaterals, such as a square, which have some sides parallel and/or some sides equal.

Parallel lines are always the same distance apart. You can check whether a pair of lines are parallel by placing a set square on one line, then placing a ruler along another edge of the set square. Hold the ruler firmly and slide the set square along the ruler. If the first edge lies along the second line, the lines are parallel.

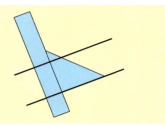

Exercise 14b

If you are not sure whether two lines are equal, *measure them*.

If you are not sure whether two lines are parallel, *check them*.

1 The Square

A(3, 2), B(11, 2), C(11, 10) and D(3, 10) are the four corners of a square. Mark these points on your own set of axes and then draw the square ABCD.

a Write down, as a number of sides of grid squares, the lengths of the sides AB, BC, CD and DA.

b Which side is parallel to AB? Are BC and AD parallel?

c What is the size of each angle of the square?

2 The Rectangle

A(2, 2), B(2, 7), C(14, 7) and D(14, 2) are the vertices of a rectangle ABCD. Draw the rectangle ABCD on your own set of axes.

a Write down the sides which are equal in length.

b Write down the pairs of sides which are parallel.

c What is the size of each angle of the rectangle?

3 The Rhombus

A(8, 1), B(11, 7), C(8, 13) and D(5, 7) are the vertices of a rhombus ABCD. Draw the rhombus on your own set of axes.

a Write down the sides which are equal in length.

b Write down the pairs of sides which are parallel.

c Measure the angles of the rhombus. Are any of the angles equal?

4 The Parallelogram

A(2, 2), B(14, 2), C(17, 7) and D(5, 7) are the vertices of a parallelogram. Draw the parallelogram on your own set of axes.

a Write down which sides are equal in length.

b Write down which sides are parallel.

c Measure the angles of the parallelogram. Write down which, if any, of the angles are equal.

5 The Trapezium

A(1, 1), B(12, 1), C(10, 5) and D(5, 5) are the vertices of a trapezium. Draw the trapezium on your own set of axes.
a   Write down which, if any, of the sides are the same length.
b   Write down which, if any, of the sides are parallel.
c   Write down which, if any, of the angles are equal.

Properties of the sides and angles of the special quadrilaterals

We can summarise our investigations in the last exercise as follows:

In a square	all four sides are the same length
	both pairs of opposite sides are parallel
	all four angles are right angles.
In a rectangle	both pairs of opposite sides are the same length
	both pairs of opposite sides are parallel
	all four angles are right angles
In a rhombus	all four sides are the same length
	both pairs of opposite sides are parallel
	the opposite angles are equal.
In a parallelogram	the opposite sides are the same length
	the opposite sides are parallel
	the opposite angles are equal.
In a trapezium	just one pair of opposite sides are parallel.

Exercise 14c

In the following questions the points A, B, C and D are the vertices of a quadrilateral. Draw the figure ABCD on your own set of axes and write down which type of quadrilateral it is.

1   A(2, 4)   B(7, 4)   C(8, 7)   D(3, 7)

2   A(2, 2)   B(6, 0)   C(7, 2)   D(3, 4)

3   A(2, 2)   B(7, 2)   C(5, 5)   D(3, 5)

4   A(2, 0)   B(6, 0)   C(6, 4)   D(2, 4)

5   A(1, 1)   B(4, 0)   C(4, 6)   D(1, 3)

6   A(3, 1)   B(6, 3)   C(3, 5)   D(0, 3)

7   A(1, 3)   B(4, 1)   C(6, 4)   D(3, 6)

8   A(2, 4)   B(3, 7)   C(9, 5)   D(8, 2)

9   A(3, 1)   B(5, 1)   C(3, 5)   D(1, 5)

10   A(0, 0)   B(5, 0)   C(8, 4)   D(3, 4)

Did you know?

René Descartes (1596–1650) was the founder of Cartesian geometry. The ideas in this chapter are based on his work.

He was born into a wealthy family and his mother died shortly after he was born. He was of frail health and was pampered.

He studied law, but decided upon a military career instead. As a gentleman soldier he had ample time to explore his mathematical ideas which seemed to come to him while he was in bed.

In this chapter you have seen that...

✔ you can write down the coordinates of a point as an ordered pair of numbers

✔ the first number (x-coordinate) gives distance across and the second number (y-coordinate) gives distance up or down

✔ there are five special quadrilaterals – square, rectangle, rhombus, parallelogram and trapezium – and you can identify them by their properties.

REVIEW TEST 2: CHAPTERS 8–14

In questions **1** to **13**, choose the letter for the correct answer.

1 At seven o'clock, the obtuse angle between the hands of a clock measures

  A 90°   **B** 120°
  C 150°   **D** 210°

2 Which of the following statements is true?
   i A reflex angle is less than 180°.
  ii Angles on a straight line add up to 180°.
 iii An angle of 180° is called a right angle.

  A None   **B** i only
  C ii only   **D** iii only

3 How many lines of symmetry does a rectangle have?

  A 1   **B** 2
  C 3   **D** 4

4 Which of the following points is on the *x*-axis?

  A (3, 0)   **B** (0, 4)
  C (3, 3)   **D** (4, 4)

5 OPQR is a rectangle. The coordinates of O, P and Q respectively are (0, 0), (7, 0) and (7, 3). What are the coordinates of R?

  A (7, 7)   **B** (0, 7)
  C (3, 7)   **D** (0, 3)

Use the figure below to answer questions **6** and **7**.

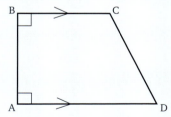

6 The line BC is

  A a ray   **B** a line segment
  C parallel to AB   **D** perpendicular to CD

7 The line AB is

   A parallel to CD   **B** perpendicular to CD

   C perpendicular to BC   **D** parallel to BC

8 $12\frac{1}{2}$ % as a fraction is

   A $\frac{1}{12}$ **B** $\frac{1}{10}$

   C $\frac{1}{8}$ **D** $\frac{1}{4}$

9 150 grams as a fraction of 1 kilogram is

   A $\frac{1}{1000}$ **B** $\frac{3}{20}$

   C $\frac{5}{20}$ **D** $\frac{15}{20}$

10 The degree of rotational symmetry of this figure is

   A 2   **B** 4

   C 6   **D** 8

11 What percentage of 14 is 42?

   A $\frac{1}{3}$%   **B** 3%

   C $33\frac{1}{3}$%   **D** 300%

12 $\frac{4}{25}$ as a percentage is

   A 0.16%   **B** 4%

   C 16%   **D** 160%

13 In a sale, all marked prices are reduced by 10%. The reduction on an item marked $2500 is

   A $100   **B** $200

   C $250   **D** $2250

14 Draw a line segment joining the points (12, 4) and (6, 2). Find, by measurement, the midpoint of the line segment, and state its coordinates.

On the same diagram draw the line segment joining (0, 4) and (18, 0). At what point do the two lines cut?

15 This figure has three lines of symmetry.
AB is 6 cm long. How long is BC?

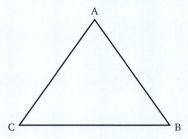

16 Write down the size of angle *a* and angle *b*.

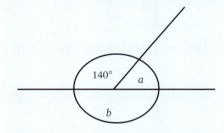

17 One jar of apricot jam weighs 950 g and costs $300.
Another jar of apricot jam weighs 300 g and costs $120.
Which jar is the better buy? Give a reason for your answer.

18 **a** A calendar is 0.375 cm thick when new. Each month is printed on a sheet 0.03125 cm thick.
What thickness remains when the calendar shows the month of April?
b Given that 1 kg ≈ 2.2 lb, calculate which is cheaper, 3 kg of meat for $1650 or 5 lb for $1800? (show all working)

15 Area

Did you know?

The original definition of a metre was the length of a platinum iridium bar kept in controlled conditions in Paris, but in 1960 it was redefined as the length of the path travelled by light in a vacuum in an interval of $\frac{1}{299\,792\,458}$ of a second.

You need to know...

✔ how to multiply and divide by 10, 100, 1000, ...
✔ the multiplication tables up to 10×10
✔ how to multiply fractions and decimals
✔ how to add and subtract decimals
✔ the properties of squares and rectangles

area, perimeter, rectangle, rectilinear, square, square centimetre,
square kilometre, square metre, square millimetre

Perimeter

The perimeter of a shape is the total distance around the edge of the shape.

The perimeter of a simple shape can be found from a diagram
that is not full size if the measurements are given.

The perimeter of this square is

$\qquad$ 4 cm + 4 cm + 4 cm + 4 cm = 16 cm

This diagram shows a metal plate for strengthening
part of a piece of furniture.
It is not drawn full size and it is not drawn to scale
but it can be used to find the perimeter of the actual shape.

To find the perimeter we need to find the missing measurements.
The length of the base line is 2 cm + 2.5 cm = 4.5 cm
and the length of the other unmarked line is also 2 cm + 2.5 cm = 4.5 cm.
Working from the top left-hand corner the perimeter of the L-shape
is therefore

$\qquad$ 2 cm + 2 cm + 2.5 cm + 2.5 cm + 4.5 cm + 4.5 cm = 18 cm

Exercise 15a

1 a Find the perimeter of this square field.
  b Fencing is sold in rolls 30 metres long. How many
    rolls of fencing are needed to go round the edge
    of this field?

2 a Find the perimeter of this rectangular
    piece of wood.
  b A finishing strip of veneer is stuck round the
    edge of this piece of wood. Veneer is sold in
    strips 1.5 metres long.
    How many strips are needed?

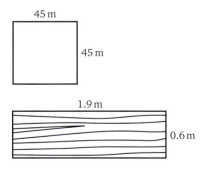

3 A square table cloth has a side of length 150 cm.
  What is the perimeter of the cloth?

4   A rectangular paving slab measures 1 metre by 1.5 metres.
    Find the perimeter of the slab.

5   A cushion cover measures 35 cm by 35 cm.

    a   What is the name of the shape of the
        cushion cover?

    b   The cover is made by sewing together two
        pieces of material. How long is the seam?

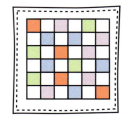

In questions **6** to **11** the shapes are all drawn on 5 mm squared
paper. Find the perimeter of each shape.

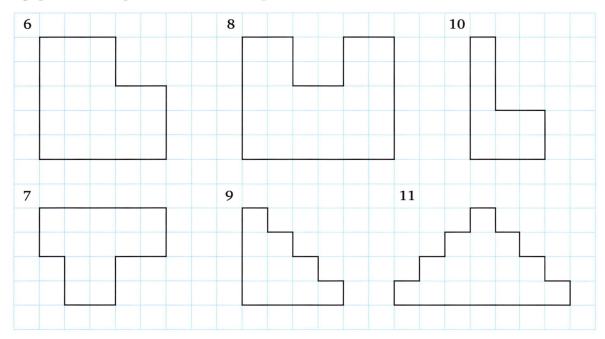

12   The following table gives some of the measurements for various
     rectangles. Find the missing values.

	Length	Breadth	Perimeter
a	4 cm		12 cm
b	5 cm		14 cm
c		3 m	16 m
d		6 mm	30 mm
e	4.6 cm	5.9 cm	
f	8.2 cm		24 cm
g		3.7 m	20 m

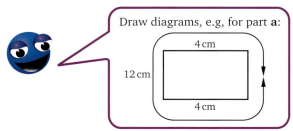

Draw diagrams, e.g, for part **a**:

Estimate the perimeter of this quadrilateral.
Now find its exact perimeter.

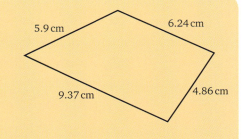

Rounding the length of each side to the nearest
whole number gives the lengths of the sides as
6 cm, 6 cm, 5 cm and 9 cm.
Estimated perimeter is 6 cm + 6 cm + 5 cm + 9 cm
$$= 26\,\text{cm}$$

Accurate value is 5.9 cm + 6.24 cm + 4.86 cm + 9.37 cm
$$= 26.37\,\text{cm}$$

13 First estimate, and then find accurately, the perimeter of each of the
following shapes.

a

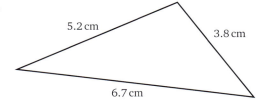

b

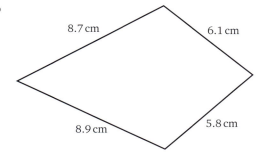

c

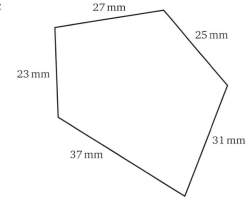

d

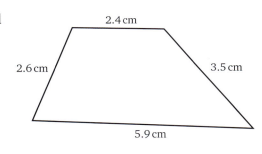

e

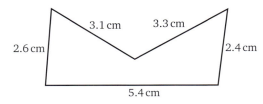

f

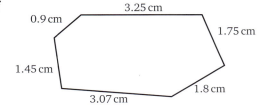

14 Find the perimeter of the following regular shapes:

    a A pentagon of side 4.6 cm.

    b A hexagon with side 3.34 cm.

    c An octagon with side 135 mm.

    d A square with side 14.7 m.

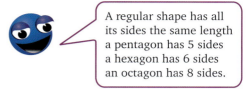

A regular shape has all its sides the same length a pentagon has 5 sides a hexagon has 6 sides an octagon has 8 sides.

Counting squares

The area of a shape or figure is the amount of surface enclosed within the lines that bound it. Below, six letters have been drawn on squared paper.

We can see by counting squares, that the area of the letter E is 15 squares.

Exercise 15b

What is the area of:

1   The letter T?   2   The letter H?

Sometimes the squares do not fit exactly on the area we are finding. When this is so we count a square if more than half of it is within the area we are finding, but exclude it if more than half of it is outside.

By counting squares in this way the approximate area of the letter A is 13 squares.

What is the approximate area of:

3   The letter P?   4   The letter O?

The next set of diagrams shows the outlines of three leaves.

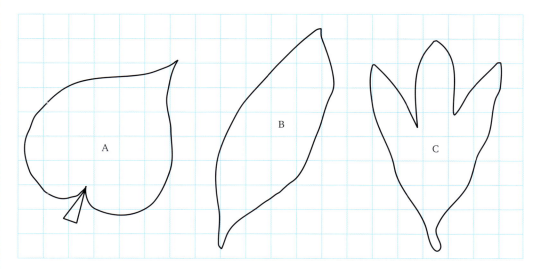

By counting squares find the approximate area of:

5   The leaf outline marked A.

6   The leaf outline marked B.

7   The leaf outline marked C.

8   Which leaf has

    a   the largest area

    b   the smallest area?

In each of the following questions find the area of the given figure by counting squares:

The following method may be used to find the area:

1.   Count the number of complete squares.

2.   Count the number of incomplete squares and divide this number by 2.

3.   Add the results obtained in 1 and 2 above.

The answer is the required area.

9

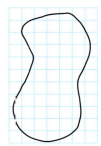

10

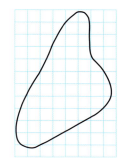

11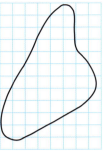

14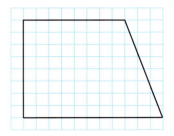

12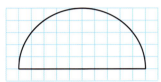

15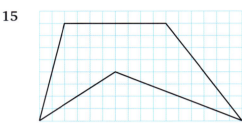

13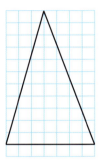

16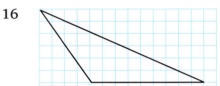

(?) Puzzle

A man has a square swimming pool in his garden with a concrete
post at the outside on each corner as shown.

He wants to double the area of his pool and still keep it square, so that
none of the posts have to be moved and are still outside the pool.

How can he do this?

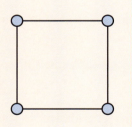

Units of area

There is nothing special about the size of square we have used. If other people
are going to understand what we are talking about when we say that the area
of a certain shape is 12 squares, we must have a square or unit of area which
everybody understands and which is always the same.

A metre is a standard length and a square with sides 1 m long is said to
have an area of one square metre. We write one square metre as $1\,m^2$.

Other agreed lengths such as millimetres, centimetres and kilometres are also in use. The unit of area used depends on what we are measuring.

We could measure the area of a small coin in square millimetres (mm²), the area of the page of a book in square centimetres (cm²), the area of a roof in square metres (m²) and the area of an island in square kilometres (km²).

Area of a square

The square is the simplest figure of which to find the area. If we have a square whose side is 4 cm long it is easy to see that we must have 16 squares, each of side 1 cm, to cover the given square:

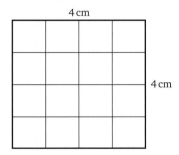

i.e. the area of a square of side 4 cm is 4×4 cm² = 16 cm².

Area of a rectangle

If we have a rectangle measuring 6 cm by 4 cm we require 4 rows each containing 6 squares of side 1 cm to cover this rectangle:

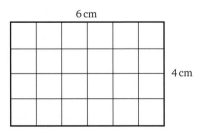

i.e. the area of the rectangle $= 6 \times 4$ cm²

$$= 24 \text{ cm}^2$$

A similar result can then be found for a rectangle of any size. For example, a rectangle of length 4 cm and breadth $2\frac{1}{2}$ cm has an area of $4 \times 2\frac{1}{2}$ cm².

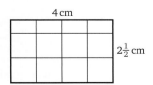

In general, for any rectangle

Area = length × breadth

Find the area of each of the following shapes, clearly stating the units involved:

1   A square of side 2 cm

2   A square of side 8 cm

3   A square of side 10 cm

4   A square of side 5 cm

5   A square of side 1.5 cm

6   A square of side 2.5 cm

7   A square of side 0.7 m

8   A square of side 1.2 cm

9   A square of side $\frac{1}{2}$ km

10 A square of side $\frac{3}{4}$ m

11 A rectangle measuring 5 cm by 6 cm

12 A rectangle measuring 6 cm by 8 cm

13 A rectangle measuring 3 m by 9 m

14 A rectangle measuring 14 cm by 20 cm

15 A rectangle measuring 1.8 mm by 2.2 mm

16 A rectangle measuring 35 km by 42 km

17 A rectangle measuring 1.5 m by 1.9 m

18 A rectangle measuring 4.8 cm by 6.3 cm

19 A rectangle measuring 95 cm by 240 cm

20 A rectangle measuring 150 mm by 240 mm

Compound figures

You can often find the area of a figure by dividing it into two or more rectangles.

Find the area of the following figure.

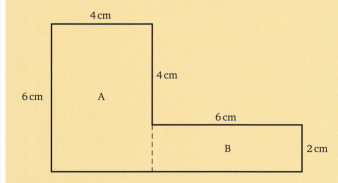

The broken line divides this shape into two rectangles, A and B.

Area of shape = area A + area B

Area of A = 6×4 cm^2 = 24 cm^2

Area of B = 6×2 cm^2 = 12 cm^2

Therefore, area of whole figure 24 cm^2 + 12 cm^2 = 36 cm^2.

Find the areas of the following figures by dividing them into rectangles.

1

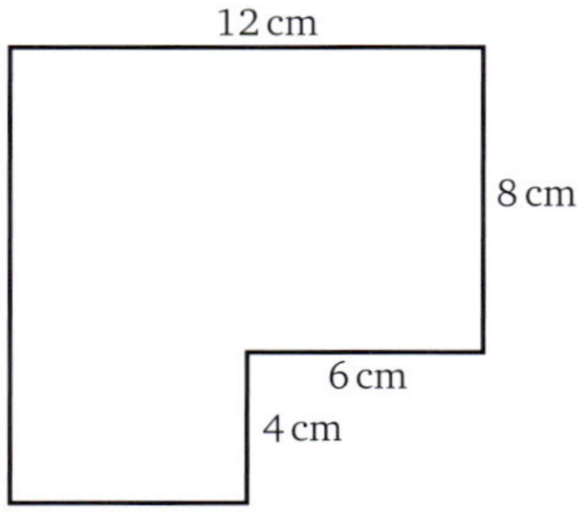

12 cm
8 cm
6 cm
4 cm

Sketch the diagram. Draw a line to divide it into two rectangles. There is often more than one way of doing this. Label the rectangles A and B. Work out and mark in any extra lengths you need to find the areas of the rectangles.

2

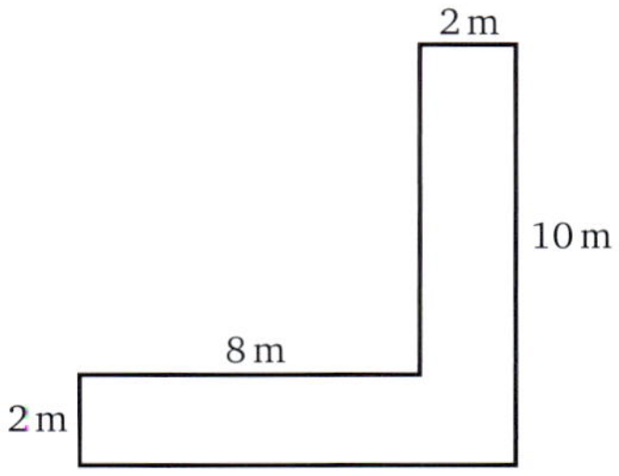

2 m
10 m
8 m
2 m

7

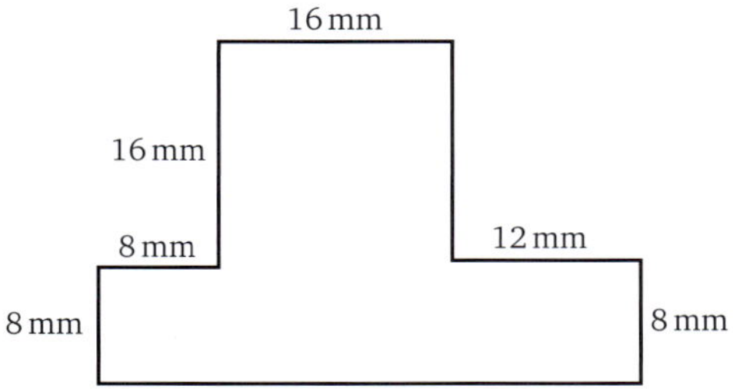

16 mm
16 mm
8 mm
12 mm
8 mm
8 mm

3

3 m
3 m
10 m
14 m

8

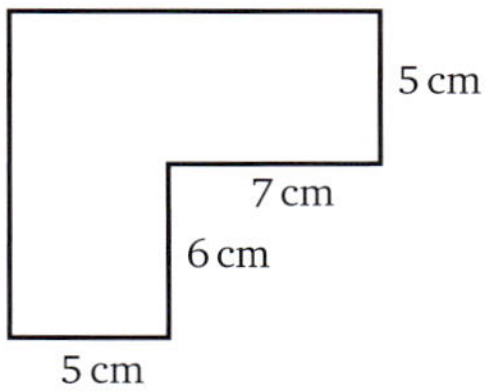

5 cm
7 cm
6 cm
5 cm

4

8 mm
24 mm
4 mm
14 mm

9

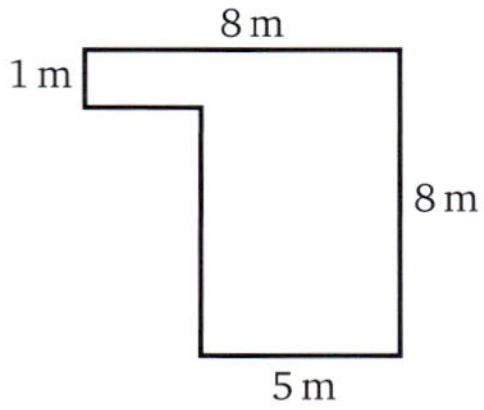

8 m
1 m
8 m
5 m

5

2 m
5 m
6 m
6 m
3 m

10

24 cm
10 cm
8 cm
18 cm

6

8 cm
3 cm
3 cm
5 cm
3 cm

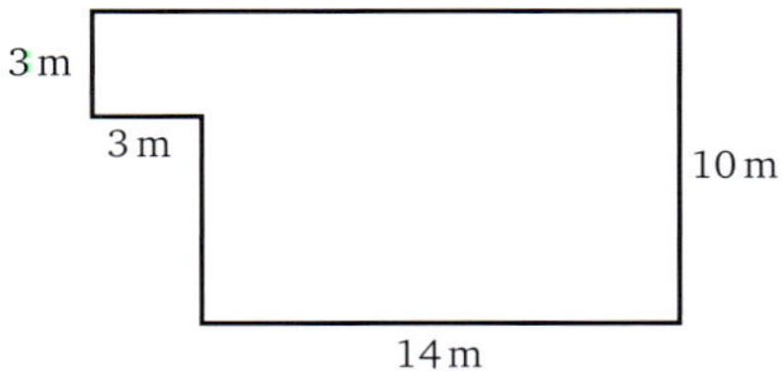

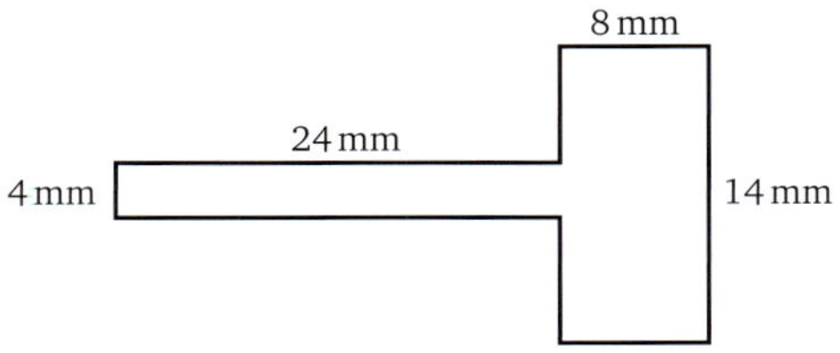

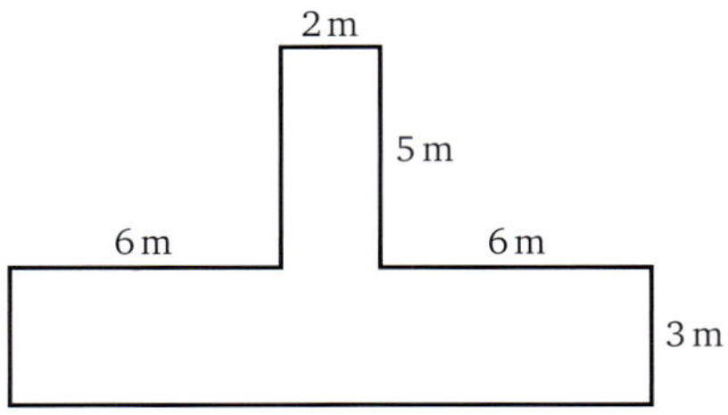

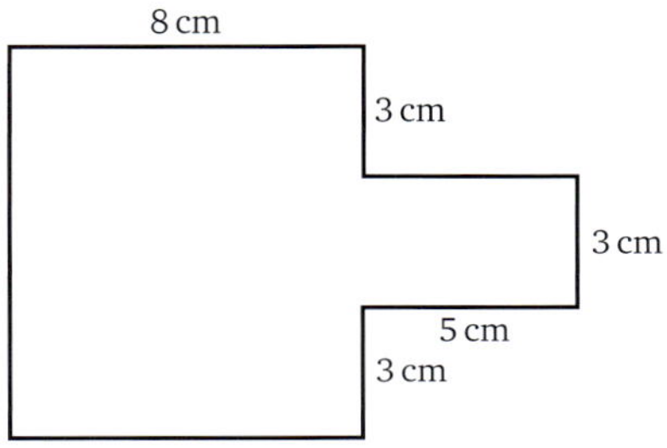

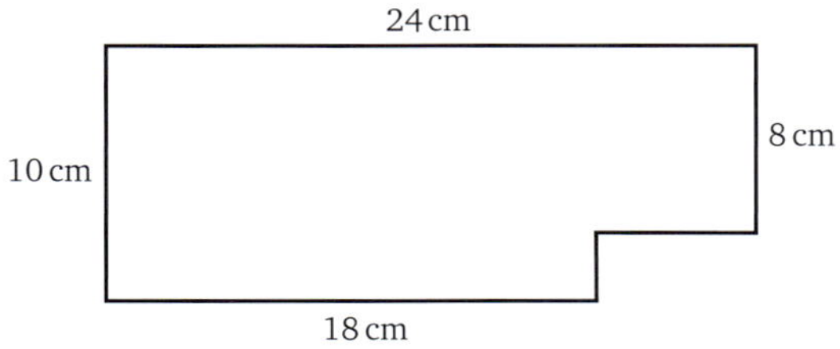

Puzzle

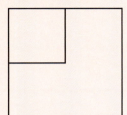

A farmer has a square field. He has already planted a quarter of the field with sugar cane as shown in the diagram. He now wants to divide the remainder of the field into four equal plots all the same size and shape. How will he do it?

Problems

Exercise 15e

For this figure, find
a the perimeter   **b** the area.

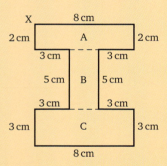

a Starting at X, the distance all round the figure and back to X is

$8 + 2 + 3 + 5 + 3 + 3 + 8 + 3 + 3 + 5 + 3 + 2$ cm $= 48$ cm.

Therefore the perimeter is 48 cm.

b Divide the figure into three rectangles A, B and C.

Then the area of A $= (8 \times 2)$ cm$^2 = 16$ cm^2

the area of B $= (5 \times 2)$ cm$^2 = 10$ cm^2

and   the area of C $= (8 \times 3)$ cm$^2 = 24$ cm^2

Therefore, the total area $= (16 + 10 + 24)$ cm$^2 = 50$ cm^2.

For each of the following figures, find   **a**   the perimeter   **b**   the area.

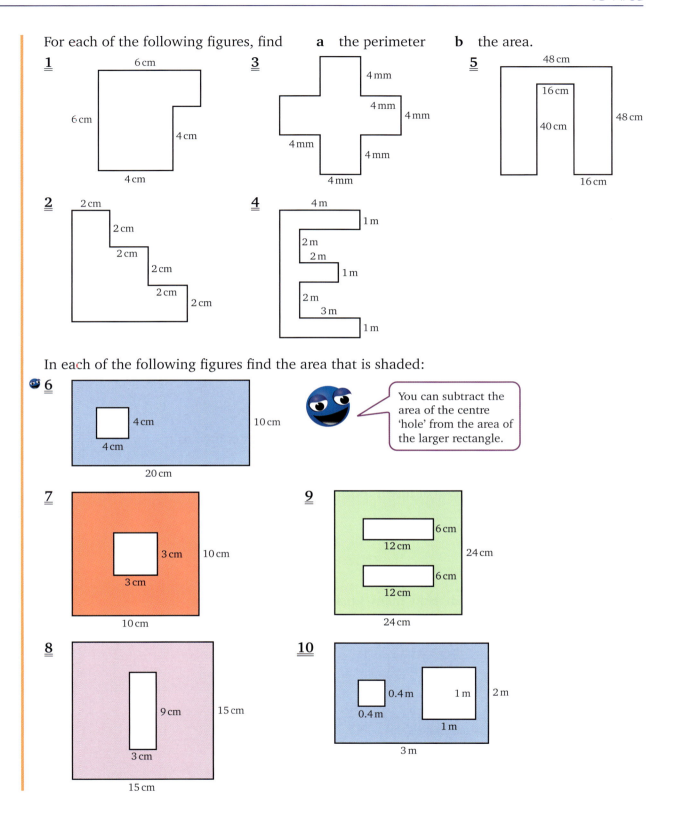

In each of the following figures find the area that is shaded:

You can subtract the area of the centre 'hole' from the area of the larger rectangle.

Investigation

Shapes made with 1 cm squares

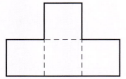

1   This shape has a perimeter of 10 cm. What is its area?

2   Find other shapes made from 1 cm squares that also have a
    perimeter of 10 cm.
    Which one has the largest area?

3   Find other shapes that have the same area as the shape above.
    Which shape has the shortest perimeter?

4   Investigate different shapes with a perimeter of 16 cm.
    Find the shape with the largest possible area.

5   Investigate different shapes with an area of 6 cm².
    Which shape has the shortest perimeter?

6   For a given area, what shape has the shortest perimeter?

7   A rectangle has the same number of square centimetres of area as it
    has centimetres of perimeter.
    Find possible whole number values for the length and breadth of this
    rectangle. (There are two different rectangles with this property.)

Exercise 15f

Draw a square of side 6 cm. How many squares of side 2 cm are required to
cover it?

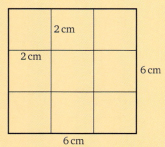

Three 2 cm squares will fit along each edge.

We see that 9 squares of side 2 cm are required to cover the larger square
whose side is 6 cm.

1 Draw a square of side 4 cm. How many squares of side 2 cm are required to cover it?

2 Draw a square of side 9 cm. How many squares of side 3 cm are required to cover it?

3 Draw a rectangle measuring 6 cm by 4 cm. How many squares of side 2 cm are required to cover it?

4 Draw a rectangle measuring 9 cm by 6 cm. How many squares of side 3 cm are required to cover it?

5 How many squares of side 5 cm are required to cover a rectangle measuring 45 cm by 25 cm?

6 How many squares of side 4 cm are required to cover a rectangle measuring 1 m by 80 cm?

Changing units of area

A square of side 1 cm may be divided into 100 equal squares of side 1 mm,

i.e. $1 \text{ cm}^2 = 100 \text{ mm}^2$

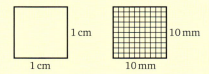

Similarly, since $1 \text{ m} = 100 \text{ cm}$

$1 \text{ square metre} = 100 \times 100 \text{ square centimetres}$

i.e. $1 \text{ m}^2 = 10\,000 \text{ cm}^2$

and as $1 \text{ km} = 1000 \text{ m}$

$1 \text{ km}^2 = 1000 \times 1000 \text{ m}^2$

i.e. $1 \text{ km}^2 = 1\,000\,000 \text{ m}^2$

When we convert from a unit of area which is large to a unit of area which is smaller we must remember that the number of units will be bigger:

e.g. $2 \text{ km}^2 = 2 \times 1\,000\,000 \text{ m}^2$

$= 2\,000\,000 \text{ m}^2$

and $12 \text{ m}^2 = 12 \times 10\,000 \text{ cm}^2$

$= 120\,000 \text{ cm}^2$

While if we convert from a unit of area which is small into one which is larger the number of units will be smaller:

e.g. $500 \text{ mm}^2 = \frac{500}{100} \text{ cm}^2$

$= 5 \text{ cm}^2$

Exercise 15g

Express $5\,m^2$ in   **a** cm^2 **b** mm^2.

a Since $1\,m^2 = 100 \times 100\,cm^2$

                  $5\,m^2 = 5 \times 100 \times 100\,cm^2$

                      $= 50\,000\,cm^2$

b Since $1\,cm^2 = 100\,mm^2$

        $50\,000\,cm^2 = 50\,000 \times 100\,mm^2 = 5\,000\,000\,mm^2$

      Therefore, $5\,m^2 = 50\,000\,cm^2 = 5\,000\,000\,mm^2$.

1   Express in cm^2:

    a $3\,m^2$ **b** $12\,m^2$ **c** $7.5\,m^2$ **d** $82\,m^2$ **e** $8\frac{1}{2}\,m^2$

2   Express in mm^2:

    a $14\,cm^2$ **b** $3\,cm^2$ **c** $7.5\,cm^2$ **d** $26\,cm^2$ **e** $32\frac{1}{2}\,cm^2$

3   Express $0.056\,m^2$ in   **a** cm^2 **b** mm^2

Express $354\,000\,000\,mm^2$ in   **a** cm^2 **b** m^2.

a Since $100\,mm^2 = 1\,cm^2$

        $354\,000\,000\,mm^2 = \dfrac{354\,000\,000}{100}\,cm^2$

                     $= 3\,540\,000\,cm^2$

b Since $100 \times 100\,cm^2 = 1\,m^2$

        $3\,540\,000\,cm^2 = \dfrac{3\,540\,000}{100 \times 100}\,m^2$

                  $= 354\,m^2$

      Therefore $354\,000\,000\,mm^2 = 3\,540\,000\,cm^2 = 354\,m^2$.

4   Express in cm^2:

    a $400\,mm^2$ **b** $2500\,mm^2$ **c** $50\,mm^2$ **d** $25\,mm^2$ **e** $734\,mm^2$

5   Express in m^2:

    a $5500\,cm^2$ **b** $140\,000\,cm^2$ **c** $760\,cm^2$ **d** $18\,600\,cm^2$ **e** $29\,700\,000\,cm^2$

6   Express in km^2:

    a $7\,500\,000\,m^2$ **b** $430\,000\,m^2$ **c** $50\,000\,m^2$ **d** $245\,000\,m^2$ **e** $176\,000\,000\,m^2$

Sometimes questions ask us to find the area of a rectangle in different square units from those in which the length and breadth are given. When this is so, we must change the units of the measurements we are given so that they 'match' the square units required in the answer.

Exercise 15h

Find the area of a rectangle measuring 50 cm by 35 cm. Give your answer in m².

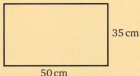

35 cm

50 cm

(Since the answer is to be given in m² we express both the length and breadth in m.)

Breadth of rectangle = 35 cm = 0.35 m

Length of rectangle = 50 cm = 0.5 m

Therefore   area of rectangle = 0.35 × 0.5 m²

= 0.175 m²

Find the area of each of the following rectangles, giving your answer in the unit in brackets:

	Length	Breadth			Length	Breadth	
1	10 m	0.5 m	(cm²)	**6**	3 m	$\frac{1}{2}$ m	(cm²)
2	6 cm	3 cm	(mm²)	**7**	$2\frac{1}{2}$ m	$1\frac{1}{2}$ m	(cm²)
3	50 m	0.35 m	(cm²)	**8**	1.5 cm	1.2 cm	(mm²)
4	1.4 m	1 m	(cm²)	**9**	0.4 km	0.3 km	(m²)
5	400 cm	200 cm	(m²)	**10**	0.45 km	0.05 km	(m²)

Mixed problems

Exercise 15i

In questions **1** to **4** find
a   the perimeter of the playing surface
b   the area of the playing surface.

1   A soccer field measuring 110 m by 75 m.

2   A rugby pitch measuring 100 m by 70 m.

3   A cricket field measuring 120 m by 70 m.

4   A tennis court measuring 26 m by 12 m.

5   A roll of wallpaper is 10 m long and 50 cm wide. Find its area in square metres.

6   A school hall measuring 20 m by 15 m is to be covered with square floor tiles of side 50 cm. How many tiles are required?

7   A rectangular carpet measures 4 m by 3 m. Find its area. How much would it cost to clean at $75 per square metre?

8   The top of my desk is 150 cm long and 60 cm wide. Find its area.

9   How many square linen serviettes, of side 50 cm, may be cut from a roll of linen 25 m long and 1 m wide?

10   How many square concrete paving slabs, each of side $\frac{3}{4}$ m, are required to pave a rectangular yard measuring 9 m by 6 m?

In this chapter you have seen that...

✔ the perimeter of a figure is the total length of all its sides

✔ you can find the area of an irregular shape by putting it on a grid and counting squares

✔ the unit of area used has to be a standard size square. Those in common use are

- square millimetres (mm^2)
- square centimetres (cm^2)
- square metres (m^2)
- square kilometres (km^2)

The relationships between them are

- $1 cm^2 = 100 mm^2$
- $1 m^2 = 10\,000 cm^2$
- $1 km^2 = 1\,000\,000 m^2$

✔ the area of a square is found by multiplying the length of a side by itself

✔ the area of a rectangle is found by multiplying its length by its breadth

✔ compound shapes can often be divided into two or more rectangles

✔ when you convert to a smaller unit of area, you multiply, and when you convert to a larger unit of area you divide.

16 Solids

At the end of this chapter you should be able to...

1 Recognise different shapes in the environment.

2 Recognise volume as a measure of space.

3 Construct a net to make a cube or cuboid.

Did you know?

Archimedes, one of the greatest mathematicians ever, lived on the island of Sicily during the 3rd century BCE. While taking a bath he discovered a law about things floating in water. He saw how this law could help the King, who thought that the man who made his crown had cheated him. Archimedes law could help to decide whether the King's crown was pure gold or not.

He was so excited that he ran out of the house naked shouting 'Eureka', which means 'I have found it'.

You need to know...

✔ how to draw accurate diagrams on squared paper
✔ how to use a protractor to draw angles

Key words

capacity, cube, cuboid, cylinder, hemisphere, net, prism, octagon, pyramid, rectilinear solid, rigid, sphere, tetrahedron, volume.

Solids

A solid is any object that takes up space, i.e. any three-dimensional object.
A solid shape must be rigid but it can be hollow.
A liquid or a gas is not a solid though it may be contained within a solid.

Solid shapes are around us everywhere in the environment.

Many of these have special names, e.g. cube, cuboid, cylinder, sphere, cone, prism and pyramid.

Some solids, like a cube, cuboid, prism and pyramid are bounded by plane surfaces and straight lines. These are called rectilinear solids.

In everyday life things we see and use approximate to one, or a combination of, some of these shapes.

 This die is approximately a cube and this brick is roughly a cuboid.

A lot of our food comes in cans that look like cylinders and a ball is usually a sphere.

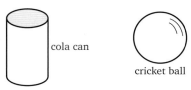

cola can

cricket ball

These are the most common basic shapes we learn about in mathematics. You are probably familiar with them already.

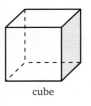

cube

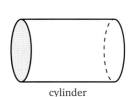

cuboid

cylinder

triangular prism

cone

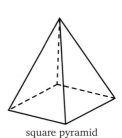

square pyramid

sphere

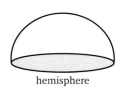

hemisphere

Exercise 16a

1   Pair each of the following objects with one or more of the shapes shown on page 238.

   a   a cricket ball
   b   a die
   c   a cornet you might buy ice cream in
   d   a cup
   e   the dome on the top of a building
   f   a tall block of apartments
   g   a book
   h   a coin
   i   a box for a DVD
   j   a wheel
   k   an orange
   l   a broom handle
   m a bus
   n   a fence post
   p   a compact disc
   q   a pencil
   r   a wardrobe
   s   a drum

2   How many of the shapes drawn above have you seen used to contain drinks?

3   Can you name an everyday object that combines two or more of these shapes?

4   In the Egyptian desert, quite near the capital Cairo, are three massive structures built by the ancient Egyptians. What mathematical name do we give to them?

5   A solid made from two or more of the basic shapes is called a compound shape. Some shapes are used more than once. Write down the basic shapes that make each of these solids.

a 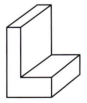   **b** 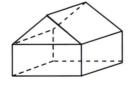   **c**

d   **e**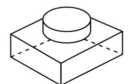

f    **g**

Volume and capacity

In the science laboratory you may well have seen a container with a spout similar to the one shown in the diagram. (Some people call this a Eureka can – do you know why?)

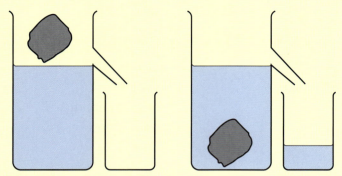

The container is filled with water to the level of the spout. Any solid that is put into the water will force a quantity of water into the measuring jug. The volume of this water will be equal to the volume of the solid. The volume of a solid is the amount of space it occupies.

Volume measures the amount of space occupied by a solid.
Volume is measured in standard units.
These are the cubic millimetre (mm^3), the cubic centimetre (cm^3) and the cubic metre (m^3).

Capacity is another measure of volume. It is an amount that can be contained within a given space and is usually used for measuring volumes of liquid and gas. The metric units of capacity are the litre and the millilitre (ml). (In the USA, the units for capacity are the pint and the gallon.)

If you open a can of soup and tip it into a rectangular food container that it just fills, you know that the capacity of the food container is exactly the same as the capacity of the can.

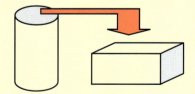

Nets

Any solid with flat faces can be made from a flat sheet.

A cube can be made from six separate squares.

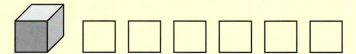

We can avoid a lot of unnecessary sticking if we join some squares together before cutting out.

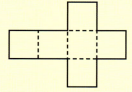

This is called a net.

There are other arrangements of six squares that can be folded up to make a cube. Not all arrangements of six squares will work however, as we will see in the next exercise.

Exercise 16b

1 Below is the net of a cube of edge 4 cm.

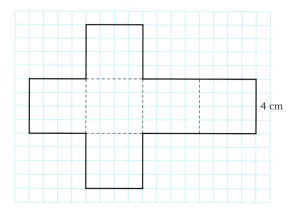

4 cm

Draw the net on 1 cm squared paper and cut it out. Fold it along the broken lines. Fix it together with sticky tape.

If you mark the faces with the numbers 1 to 6, you can make a die.

2 Draw this net full-size on 1 cm squared paper.

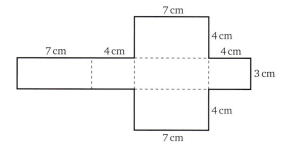

7 cm

7 cm   4 cm   4 cm

4 cm   3 cm

4 cm

7 cm

Cut the net out and fold along the dotted lines. Stick the edges together.

Using your cuboid, answer the following:

a **i** How many faces are rectangles measuring 7 cm by 4 cm?

    ii How many faces are rectangles measuring 7 cm by 3 cm?

    iii What are the measurements of the remaining faces?

b Draw another arrangement of the rectangles that will fold up to make this cuboid.

3 This cuboid is 4 cm long, 2 cm wide and 1 cm high.

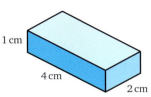

a How many faces does this cuboid have?

b Sketch the faces, showing their measurements.

c On 1 cm squared paper, draw a net that will make this cuboid.

4 This net will make a cuboid.

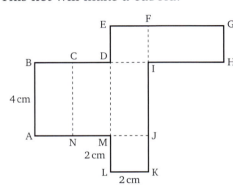

a Sketch the cuboid, and show its measurements.

b Which edge joins HI?

c Which corners meet at A?

5 This cube is cut along the edges drawn with a coloured line and flattened out.

Draw the flattened shape.

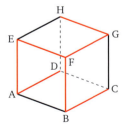

6 Here are two arrangements of six squares.

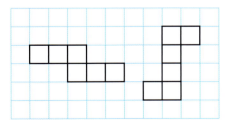

a Copy these on 1 cm squared paper.

b Draw as many other arrangements of six squares as you can find.

c Which of your arrangements, including the two given here, will fold up to make a cube? If you cannot tell by looking, cut them out and try to make a cube.

7

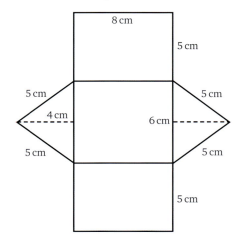

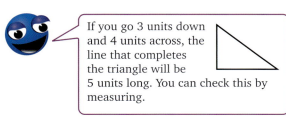

If you go 3 units down and 4 units across, the line that completes the triangle will be 5 units long. You can check this by measuring.

Draw this net on 1 cm squared paper. Cut it out and fold it to make a solid.

You have made a triangular prism similar to one of the basic solids shown at the beginning of the chapter.

8 Draw this net on 1 cm squared paper.
Cut it out and fold the triangles up to meet at a point.

The solid you have made is called a square pyramid similar to one of the basic solids shown at the beginning of the chapter.

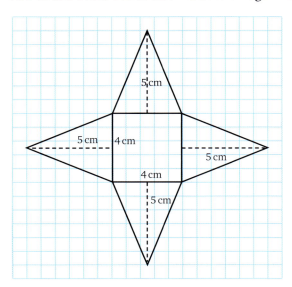

9 This is a net for a solid with four identical faces.

Each face is an equilateral triangle – a triangle that has all its sides and angles equal.

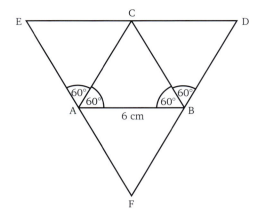

Draw this net on plain paper. Start with AB = 6 cm.

Now use your protractor to mark angle ABC = 60° and angle BAC = 60°.

The two lines you have drawn to form the angles meet at C.

In a similar way draw the other triangles.

After cutting out triangle DEF fold the three outside triangles so that D, E and F meet at a point.

The solid you have made is called a tetrahedron.

? Puzzle

Everton is returning from the local farm with an 8-litre can which is full of milk. He meets Amy who is going to the same farm for milk. She has an empty 5-litre can and an empty 3-litre can. Everton knows that there is no milk left at the farm, so being the kind boy he is, decides to share his milk equally with Amy. How do they do it using only the three containers they have?

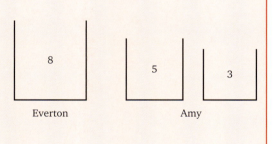

Practical work

Have you ever heard of a Moebius strip? It is named after the German mathematician August Ferdinand Moebius and has a very unusual property. Take a strip of paper, twist it once and join the ends together. You started with a sheet that had two surfaces but the Moebius strip has only one surface. What is more it has only one edge. Try making one. Then draw a line along its surface – what happens?

In this chapter you have seen...

✔ how to recognise some of the basic solids

✔ that some everyday solids are made up of two or more of the basic shapes

✔ that volume and capacity are measures of the same space

✔ how to draw a net for a cube and a cuboid

✔ how other nets can be used to give a triangular prism, a square pyramid, an octagonal prism and a tetrahedron.

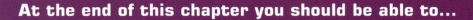

17 Reflections

At the end of this chapter you should be able to...

1 Understand mathematical reflection.

2 Reflect an object in a mirror line.

3 Find the mirror line given the object and its image.

4 Identify a reflection.

You need to know...

✔ how to plot points on a set of x and y axes

✔ the meaning of symmetry

Key words

axis of symmetry, corresponding vertices, image, invariant line, invariant point, line symmetry, mapped, midpoint, mirror line, object, parallel, perpendicular, perpendicular bisector, right-angle, reflection.

Reflections

Consider a piece of paper, with a drawing on it, lying on a table. Stand a mirror upright on the paper and the reflection can be seen as in the picture.

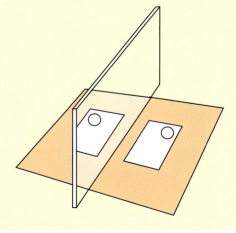

If we did not know about such things as mirrors, we might imagine that there were two pieces of paper lying on the table like this:

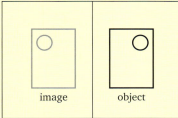

image   object

The *object* and the *image* together form a symmetrical shape and the *mirror line* is the axis of symmetry.

Exercise 17a

In this exercise it may be helpful to use a small rectangular mirror, or you can use tracing paper to trace the object and turn the tracing paper over, to find the shape of the image.

Copy the objects and mirror lines (indicated by dotted lines) on to squared paper and draw the image of each object.

1
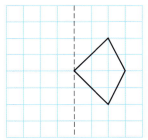

3
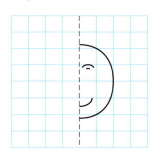

5
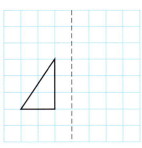

2
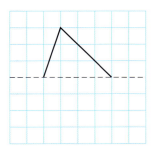

4
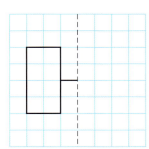

6
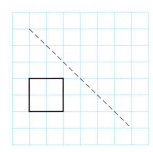

Copy triangle ABC and the mirror line on to squared paper. Draw the image.
Label the corresponding vertices (corners) of the image A′, B′, C′.

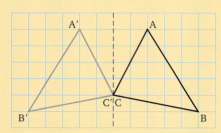

(In this case C and C′ are the same point.)

In each of the following questions, copy the object and the mirror line on to
squared paper. Draw the image. Label the vertices of the object A, B, C, etc.
and label the corresponding vertices of the image A′, B′, C′, etc.

7
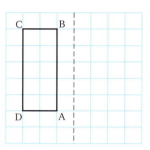

8
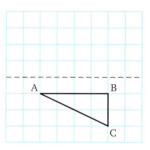

9
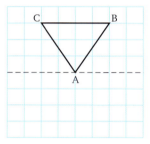

In mathematical reflection, though not in real life, the object can cross
the mirror line.

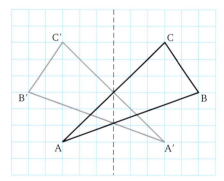

10
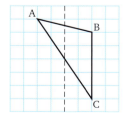

11
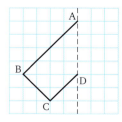

12
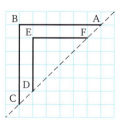

13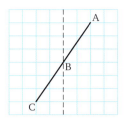

15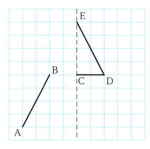

17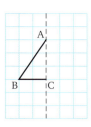

14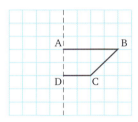

16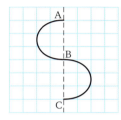

18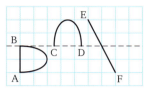

19 Which points in questions **7** to **18** are labelled twice? What is special about their positions?

20 In the diagram for question **10**, join A and A′.
   a Measure the distances of A and A′ from the mirror line. What do you notice?
   b At what angle does the line AA′ cut the mirror line?

21 Repeat question **20** on other suitable diagrams, in each case joining each object point to its image point. What conclusions do you draw?

Invariant points

A point that is its own image, i.e. such that the object point and its image are in the same place, is called an *invariant point*. The previous examples showed that, with reflection, the invariant points lie on the mirror line. The mirror line is an *invariant line*.

Finding the mirror line

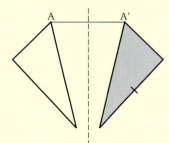

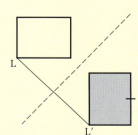

We can see from these diagrams, and from the work in the previous exercise, that the object and image points are at equal distances from the mirror line, and the lines joining them (e.g. AA′ and LL′) are perpendicular (at right angles) to the mirror line.

Find the mirror line if △A′B′C′ is the image of △ABC.

The mirror line is halfway between an object point and its image and perpendicular to the line through them.

So the mirror line is halfway between B and B′ and perpendicular to the line BB′. Check that it also goes through the midpoint of CC′.

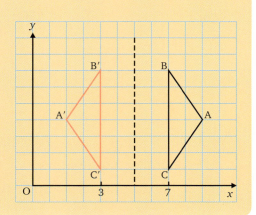

Copy the diagrams in questions **1** to **4** and draw in the mirror lines.

1

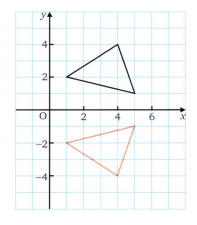

3

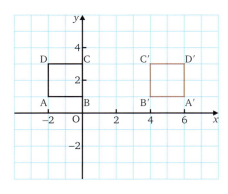

2

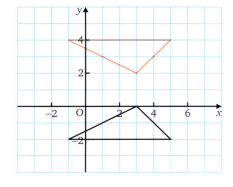

4

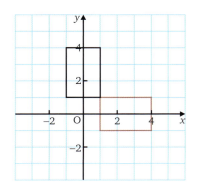

If A′B′C′ is the reflection of ABC,
draw the mirror line.

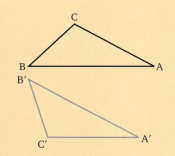

(Join AA′ and BB′ and find their midpoints,
marking them P and Q. Then PQ is the mirror line.)

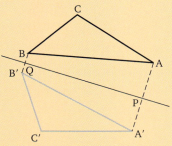

Whenever you attempt to draw a mirror line in this way, always check that
the mirror line is at right angles to AA′ and BB′. If it is not, then A′B′C′ cannot
be a reflection of ABC.

5   Trace the diagrams and draw the mirror lines.

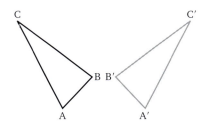

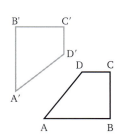

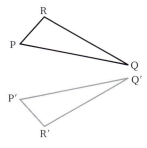

Drawing the mirror line

If we have only one point and its image, and we cannot
use squares to guide us, we can use the fact that the
mirror line goes through the midpoint of AA′ and is
perpendicular to AA′. So you can draw the mirror line.
The line through the midpoints of a line segment and
perpendicular to it is called the perpendicular bisector.

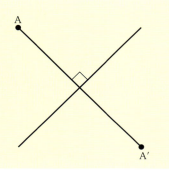

Exercise 17c

1 On plain paper mark two points P and P' about 10 cm apart in the middle of the page and draw the perpendicular bisector of PP'. Join PP' and check that it is cut in half by the line you have drawn and that the two lines cut at right angles. Are we correct in saying that P' is the reflection of P in the drawn line?

2 On squared paper draw axes for x and y from 0 to 10, using 1 cm to 1 unit. A is the point (10, 7) and A' is the point (2, 2). Draw the mirror line so that A' is the reflection of A.

3 Draw axes for x and y from 0 to 9, using 1 cm to 1 unit. B is the point (0, 1) and B' is the point (7, 4). Draw the mirror line so that B' is the reflection of B.

A reflection is only one of many ways of making an image. You will be shown some others in later books.

Exercise 17d

In the following questions, which images of △ABC are given by reflections?

1

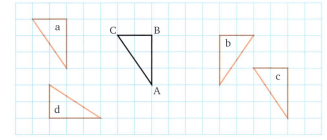

2

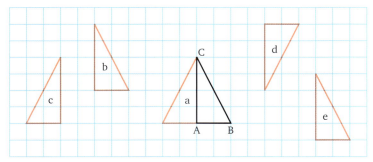

3

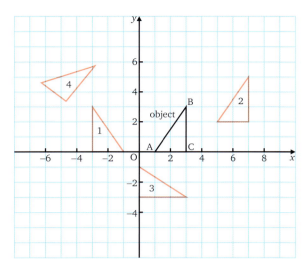

4

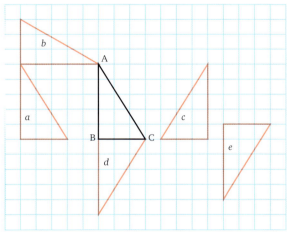

5 Which images of rectangle PQRS are reflections?

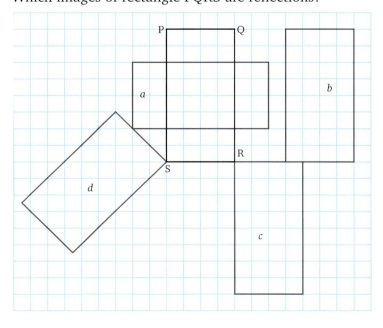

6 Which images of the shape LMNP are reflections?

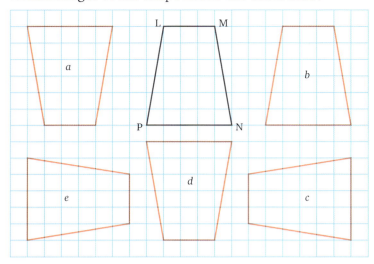

Strip patterns

Strip patterns appear in borders on walls and on textiles.

Strip patterns are made by starting with a single plane shape and then reflecting that shape.

Starting with [shape] we can reflect it to give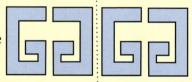

We can then reflect the pair to give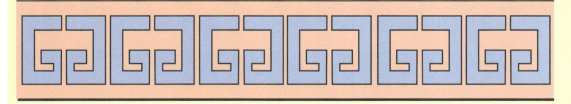

If we repeat this several times we get a strip pattern:

Any plane shape can be used and it can be reflected in a line crossing the shape.

The strip pattern below uses this basic shape, which is first reflected in the dotted line.

Then the combined shape is reflected in the dotted line below and this is repeated.

Exercise 17e

Copy these shapes on squared paper and use them to make a strip pattern using

a reflections in a line outside the shape

b reflections in a line through the shape.

1

2

3

4 Draw your own shape on squared paper and make a strip pattern with it.

5 Identify the basic pattern used to make these patterns and sketch it. There are several possible shapes.

a

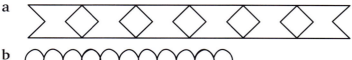

b

c

Textile patterns

We can cover a surface with a pattern by starting with a basic shape and reflecting it.

This pattern uses this shape ⌐ and reflects it in vertical lines and horizontal lines.

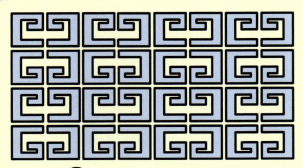

Starting with this shape (we can build up a pattern by first reflecting it in a line at 45° to the horizontal:

next reflecting the pattern in a line perpendicular to the first line:

then reflecting the pattern in a vertical line:

then reflecting the pattern in a horizontal line:

Continuing this we can build up a pattern to cover a surface.

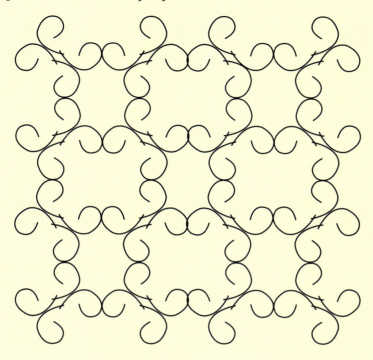

Exercise 17f

1   Copy these shapes onto squared paper and use reflections to create a
    pattern.

a

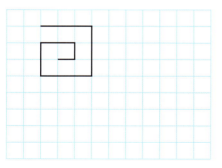

b

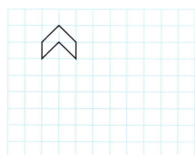

c

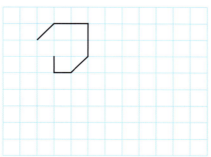

2 Use your own simple pattern to make a pattern to cover a surface.

3 This pattern is made by reflecting this shape:

Identify the mirror lines used.

⚠ Investigation

Tangrams

What is a tangram?

A tangram is a puzzle composed of seven parts called tans. The tans are formed by cutting a square and its interior into five triangles, a square and a parallelogram.

Measure all angles of the tans. What do you find?

Find the areas of a large triangle and a small triangle. What is the relationship?

Can you find other relationships?

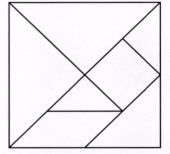

Tangram play began in China and was introduced into the western world during the 1800s. John Q. Adams and Edgar A. Poe are said to have enjoyed tangrams.

The object of the puzzle is to put the seven tans together to form outlines of all sorts. Use all the pieces, and do not overlap the tans.

It is said that tangrams contain serious as well as playful mathematics. But, what doesn't?

In this chapter you have seen that...

✔ when an object is reflected in a mirror line, the object and the image are symmetrical about the mirror line

✔ the mirror line is the perpendicular bisector of the line joining a point on the object to the corresponding point on the image.

Did you know?

Arabic numerals became known in the West through a book by the Arabian mathematician Mohammed ibn Musa al-Khowarizmi, written in the year 820 under the title *Al-jabr W'almuqabala*. It is said that the word 'algebra' came from the title of this book.

You need to know...

✔ how to work with simple numbers

Key words

equation, expression, like terms, solve, unlike terms

Algebra uses letters for numbers. These numbers may be unknown (but can sometimes be found), such as the number of cakes that a bakery needs to make to maximise profit. They may also have values that can vary, such as the lengths of rooms.

The idea of equations

'I think of a number, and take away 3; the result is 7.'

We can see the number must be 10.

Using a letter to stand for the unknown number we can write the first sentence as an equation:

$$x - 3 = 7$$

Then if $\quad x = 10$

$$10 - 3 = 7$$

so $x = 10$ fits the equation.

Exercise 18a

Form equations to illustrate the following statements and find the unknown numbers:

I think of a number, add 4 and the result is 10.

Let the number be x.

Then add 4 to x. This gives $x + 4$, which we know is 10, i.e. $x + 4$ and 10 are the same so they are equal. This gives the equation.

The equation is $x + 4 = 10$.

The number is 6.

1 I think of a number, subtract 3 and get 4.

2 I think of a number, add 1 and the result is 3.

3 If a number is added to 3 we get 9.

4 If 5 is subtracted from a number we get 2.

I think of a number, multiply it by 3 and the result is 12.

Let the number be x.

Multiplying by 3 gives $3 \times x$ which can be shortened to $3x$.

We know that $3x$ gives 12.

So the equation is $3x = 12$.

The number is 4.

5 I think of a number, double it and get 8.

6 If a number is multiplied by 7 the result is 14.

7   When we multiply a number by 3 we get 15.

8   6 times an unknown number gives 24.

Write sentences to show the meaning of the following equations:

$4x = 20$

$4x = 20$ means 4 times an unknown number gives 20, or, I think of a number, multiply it by 4 and the result is 20.

9 $3x = 18$	11 $x - 2 = 9$	13 $5 + x = 7$	15 $4x = 8$
10 $x + 6 = 7$	12 $5x = 20$	14 $x - 4 = 1$	16 $x + 1 = 4$

 Puzzle

In three years time I shall be three times as old as I was three years ago.
How old am I?

Solving equations

Some equations need an organised approach, not guesswork.

Imagine a balance:

On this side there is a bag containing an unknown number of marbles, say x marbles, and 4 loose marbles.

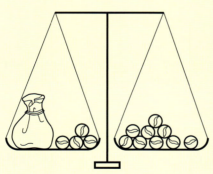

On this side, there are 9 separate marbles, balancing the marbles on the other side.

$x + 4 = 9$

Take 4 loose marbles from each side, so that the two sides still balance.

$x = 5$

We write: $x + 4 = 9$

Take 4 from both sides $x = 5$

When we have found the value of x we have *solved the equation*.

As a second example suppose that:

On this side there is a bag that originally held x marbles but now has 2 missing.

On this side, there are 5 loose marbles.

$$x - 2 = 5$$

We can make the bag complete by putting back 2 marbles but, to keep the balance, we must add 2 marbles to the right-hand side also.

So we write $x - 2 = 5$

Add 2 to both sides $x = 7$

Whatever you do to one side of an equation you must also do to the other side.

Exercise 18b

Solve the following equations:

$y + 4 = 6$

$y + 4 = 6$

Take 4 from both sides $y = 2$

1	$x + 7 = 15$	7	$a + 5 = 11$
2	$x + 9 = 18$	8	$9 + a = 15$
3	$10 + y = 12$	**9**	$a + 1 = 6$
4	$2 + c = 9$	**10**	$a + 8 = 15$
5	$a + 3 = 7$	**11**	$7 + c = 10$
6	$x + 4 = 9$	**12**	$c + 2 = 3$

You can 'see' the solution of these equations without doing any working – use this to check your answers.

$x - 6 = 2$

$x - 6 = 2$

Add 6 to both sides $x = 8$

13	$x - 6 = 4$	**16**	$x - 4 = 6$	**19**	$s - 4 = 1$	**22**	$x - 3 = 0$
14	$a - 2 = 1$	**17**	$c - 8 = 1$	**20**	$x - 9 = 3$	**23**	$c - 1 = 1$
15	$y - 3 = 5$	**18**	$x - 5 = 7$	**21**	$a - 4 = 8$	**24**	$y - 7 = 2$

Exercise 18c

Sometimes the letter term is on the right-hand side instead of the left.

$3 = x - 4$

$$3 = x - 4$$

Add 4 to both sides $\qquad 7 = x \qquad$ ($7 = x$ is the same as $x = 7$)

$$x = 7$$

Solve the following equations:

1	$4 = x + 2$	**10**	$6 + c = 10$	**19**	$x - 1 = 4$	**28**	$y - 9 = 14$
2	$6 = x - 3$	**11**	$7 = x + 3$	**20**	$10 = a - 1$	**29**	$2 = z - 2$
3	$7 = a + 4$	**12**	$x + 1 = 9$	**21**	$c - 7 = 9$	**30**	$x + 1 = 8$
4	$6 = x - 7$	**13**	$x + 3 = 15$	**22**	$x - 4 = 8$	**31**	$x - 1 = 8$
5	$1 = c - 2$	**14**	$y - 6 = 4$	**23**	$y - 1 = 9$	**32**	$x - 8 = 1$
6	$5 = s + 2$	**15**	$x - 7 = 4$	**24**	$x - 3 = 6$	**33**	$c + 5 = 9$
7	$x + 3 = 10$	**16**	$6 = x - 4$	**25**	$c - 7 = 10$	**34**	$d - 3 = 1$
8	$c + 4 = 4$	**17**	$x - 4 = 2$	**26**	$4 = b - 1$	**35**	$1 = c - 3$
9	$3 = b + 2$	**18**	$x - 9 = 2$	**27**	$x - 4 = 12$	**36**	$z + 3 = 5$

Multiples of *x*

Imagine that on this side of the scales there are 3 bags each containing an equal unknown number of marbles, say x in each.

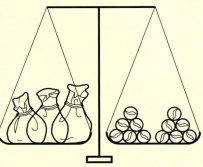

On this side there are 12 loose marbles.

$$3 \times x = 12$$
$$3x = 12$$

We can keep the balance if we divide the contents of each scale pan by 3.

$x = 4$

Exercise 18d

Solve $6x = 12$

$$6x = 12$$

Divide both sides by 6 $x = 2$

Solve $3x = 7$

$$3x = 7$$

Divide both sides by 3 $x = \frac{7}{3}$

$x = 2\frac{1}{3}$

Solve the following equations:

1	$5x = 10$	**7**	$3a = 1$	**13**	$6x = 36$	**19**	$3x = 27$
2	$3x = 9$	**8**	$6z = 18$	**14**	$6x = 6$	**20**	$8x = 16$
3	$2x = 5$	**9**	$5p = 7$	**15**	$6x = 1$	**21**	$4y = 3$
4	$7x = 21$	**10**	$2x = 40$	**16**	$5z = 10$	**22**	$5x = 6$
5	$4b = 16$	**11**	$7y = 14$	**17**	$5z = 9$	**23**	$2z = 10$
6	$4c = 9$	**12**	$6a = 3$	**18**	$2y = 7$	**24**	$7x = 1$

Mixed operations

Exercise 18e

Solve the following equations:

1	$x + 4 = 8$	**5**	$5y = 6$	**9**	$2x = 11$
2	$x - 4 = 8$	**6**	$4x = 12$	**10**	$x - 2 = 11$
3	$4x = 8$	**7**	$4 + x = 12$	**11**	$12 = x + 4$
4	$5 + y = 6$	**8**	$x - 4 = 12$	**12**	$x - 12 = 4$

13	$8 = c + 2$	**18**	$3 = a - 4$	**23**	$5a = 25$
14	$3x = 10$	**19**	$x + 3 = 5$	**24**	$a + 5 = 25$
15	$20 = 4x$	**20**	$3x = 5$	**25**	$a - 5 = 25$
16	$7y = 2$	**21**	$z - 5 = 6$	**26**	$a - 25 = 5$
17	$3x = 8$	**22**	$c + 5 = 5$	**27**	$25a = 5$

Two operations

Exercise 18f

The aim is to get the letter term on its own.

Solve $7 = 3x - 5$

$$7 = 3x - 5$$

Add 5 to both sides (to isolate the x term) $12 = 3x$

Divide both sides by 3 $4 = x$

i.e. $x = 4$

Solve $2x + 3 = 5$

$$2x + 3 = 5$$

Take 3 from both sides (to get $2x$ on its own) $2x = 2$

Divide both sides by 2 $x = 1$

(It is possible to check whether your answer is correct. We can put $x = 1$ in the left-hand side of the equation and see if we get the same value on the right-hand side.)

Check: If $x = 1$, left-hand side $= 2 \times 1 + 3 = 5$

Right-hand side $= 5$, so $x = 1$ fits the equation.

Solve the following equations:

1	$6f + 2 = 26$	**5**	$7x + 1 = 22$	**9**	$3p - 4 = 4$
2	$4x + 7 = 19$	**6**	$3a + 12 = 12$	**10**	$3x + 4 = 25$
3	$17 = 7x + 3$	**7**	$10 = 10x - 50$		
4	$4d - 5 = 19$	**8**	$6 = 2h - 4$		

Add 50 to both sides.

11	$2x + 15 = 25$	**19**	$15 = 1 + 7c$	**27**	$16 = 7x - 1$
12	$13 = 3e + 4$	**20**	$9x - 4 = 14$	**28**	$10x - 6 = 24$
13	$5z - 9 = 16$	**21**	$3x - 2 = 3$	**29**	$5x - 7 = 4$
14	$20 = 12x - 4$	**22**	$7 = 2z + 6$	**30**	$8 = 3x + 7$
15	$9g + 1 = 28$	**23**	$5 = 7x - 23$	**31**	$9 = 6a - 27$
16	$9 = 8x - 15$	**24**	$2x + 6 = 6$	**32**	$4z + 3 = 4$
17	$8 = 8 + 3z$	**25**	$19x - 16 = 22$	**33**	$2x + 4 = 14$
18	$5x - 4 = 5$	**26**	$3x + 1 = 11$	**34**	$3 = 7x - 3$

Problems

Exercise 18g

I think of a number, double it and add 3. The result is 15.

What is the number?

Let the number be x. Doubling it gives $2x$, then adding 3 gives $2x + 3$.

The result is 15, so the equation is $\qquad 2x + 3 = 15$

Take 3 from both sides $\qquad\qquad\qquad 2x = 12$

Divide both sides by 2 $\qquad\qquad\qquad x = 6$

The number is 6.

The side of a square is x cm. Its perimeter is 20 cm.

Find x.

Draw a diagram.

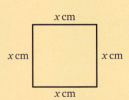

Now you can see that the perimeter is $(x + x + x + x)$ cm which is $4x$ cm.

You also know that the perimeter is 20 cm,

so the equation is $\qquad\qquad\qquad 4x = 20$

Divide both sides by 4 $\qquad\qquad\qquad x = 5$

Check: $5 + 5 + 5 + 5 = 10 + 5 + 5 = 15 + 5 = 20$.

Form equations and solve the problems:

1 I think of a number, multiply it by 4 and subtract 8. The result is 20. What is the number?

2 I think of a number, multiply it by 6 and subtract 12. The result is 30. What is the number?

3 I think of a number, multiply it by 3 and add 6. The result is 21. What is the number?

4 When 8 is added to an unknown number the result is 10. What is the number?

5 I think of a number, multiply it by 3 and add the result to 7. The total is 28. What is the number?

6 The sides of a rectangle are x cm and 3 cm. Its perimeter is 24 cm. Find x.

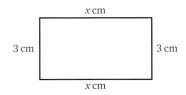

Find the perimeter from the diagram first.

7 The lengths of the three sides of a triangle are x cm, x cm and 6 cm. Its perimeter is 20 cm. Find x.

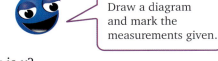

Draw a diagram and mark the measurements given.

8 Mary and Jean each have x sweets and Susan has 10 sweets. Amongst them they have 24 sweets. What is x?

9 Three boys had x sweets each. Amongst them they gave 9 sweets to a fourth boy and then found that they had 18 sweets left altogether. Find x.

10 I have two pieces of ribbon, each x cm long and a third piece 9 cm long. Altogether there are 31 cm of ribbon. What is the length of each of the first two pieces?

Did you know?

If the sum of two numbers is 24 and the difference is 6, then one of the numbers is $\frac{(24+6)}{2} = 15$

or

If the sum of two numbers is 28 and the difference is 12, then one of them is $\frac{(28+12)}{2} = 20$.

Use your algebra to prove that this is always so.

Simplifying expressions

Consider $3x + 5x - 4x + 2x$. This is called an *expression*. There is no equals sign. $3x$, $5x$, $4x$ and $2x$ are *terms* in the expression.

The terms in any expression are separated by + or − signs.

If x stands for an unknown number, then the x in each term represents the same number. So $3x$, $5x$, $4x$ and $2x$ are of the same type. They are called like terms.

$3x + 5x - 4x + 2x$ can be simplified by addition and subtraction to $6x$. This is called *collecting like terms*.

Exercise 18h

$4h - 6h + 7h - h$

You can do the addition before the subtraction, i.e. $4h + 7h - 6h - h$

$4h - 6h + 7h - h = 4h$

Simplify:

1   $3x + x + 4x + 2x$

2   $3x - x + 4x - 2x$

3   $8x - 6x$

4   $6 - 1 + 4 - 7$

5   $9y - 3y + 2y$

6   $2 - 3 + 9 - 1$

7   $5 - 3 - 1$

8   $3x - 2x - x$

Remember that the sign in front of a number applies to that number only.

Unlike terms

$3x + 2x - 7$ can be simplified to $5x - 7$, and $5x - 2y + 4x + 3y$ can be simplified to $9x + y$.

Terms containing x are different from terms without an x. They are called unlike terms and cannot be collected (you cannot take 7 away from an unknown number). Similarly $9x$ and $5y$ are unlike terms as x and y represent different numbers. Therefore $9x - 5y$ cannot be simplified.

Exercise 18i

Simplify $3x - 4 + 7 - 2x + 4x$

You can rearrange this to have the like terms together, i.e. $3x - 2x + 4x + 7 - 4$

$$3x - 4 + 7 - 2x + 4x = 5x + 3$$

Simplify $2x + 4y - x + 5y$

$$2x + 4y - x + 5y = 2x - x + 4y + 5y = x + 9y$$

Simplify:

1 $2x + 4 + 3 + 5x$

2 $2x - 4 + 3x + 9$

3 $5x - 2 + 3 - x$

4 $4a + 5c + 6a$

5 $6x + 5y + 2x - 3y$

6 $6x + 5y + 2x + 3y$

7 $6x + 5y - 2x - 3y$

8 $6x + 5y - 2x + 3y$

9 $4x + 1 + 3x + 2 + x$

10 $6x + 9 + 2x - 1$

11 $7x - 3 + 9 - 4x$

12 $9x + 3y + 10x$

13 $6x + 5y + 2x + 3y + 2x$

14 $6x + 5y - 2x - 3y + 7x + y$

15 $30x + 2 - 15x - 6 + 4$

16 $2z + 3x + 4y + 6z + x - 3y$

 There are three sets of like terms here: xs, ys and zs.

17 $4x + 3y - 4 + 6x - 2y + 7 - x$

18 $7x + 3 - 9 - 9x + 2x + 6 + 11$

19 What is the difference between an expression and an equation?

Equations with letter terms on both sides

Some equations have letter terms on both sides. Consider the equation

$$5x + 1 = 2x + 9$$

We want to have a letter term on one side only so we need to take $2x$ from both sides. This gives

$$3x + 1 = 9$$

and we can go on to solve the equation as before.

Notice that we want the letter term on the side which has the greater number of xs to start with.

If we look at the equation

$$9 - 4x = 2x + 4$$

we can see that there is a lack of xs on the left-hand side, so there are more xs on the right-hand side. Add $4x$ to both sides and then the equation becomes

$$9 = 6x + 4$$

and we can go on as before.

Exercise 18j

Deal with the letters first, then the numbers.

Solve $5x + 2 = 2x + 9$

$$5x + 2 = 2x + 9$$

$2x < 5x$ so take $2x$ from both sides ($<$ means 'less than') $\quad 3x + 2 = 9$

Take 2 from both sides $\quad\quad\quad\quad\quad\quad\quad\quad\quad\quad\quad\quad 3x = 7$

Divide both sides by 3 $\quad\quad\quad\quad\quad\quad\quad\quad\quad\quad\quad x = \dfrac{7}{3} = 2\dfrac{1}{3}$

Solve the following equations:

1	$3x + 4 = 2x + 8$		**5**	$7x + 3 = 3x + 31$
2	$x + 7 = 4x + 4$		**6**	$6z + 4 = 2z + 1$
3	$2x + 5 = 5x - 4$		**7**	$7x - 25 = 3x - 1$
4	$3x - 1 = 5x - 11$		**8**	$11x - 6 = 8x + 9$

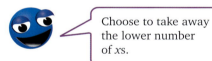

Choose to take away the lower number of xs.

Equations containing like terms

If there are a lot of terms in an equation, first collect the like terms on each side separately.

This flow chart summarises the processes we have covered for solving equations.

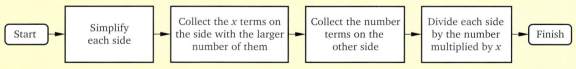

Start → Simplify each side → Collect the x terms on the side with the larger number of them → Collect the number terms on the other side → Divide each side by the number multiplied by x → Finish

Remember that we collect the x (or number) terms on one side of an equation by *adding or subtracting the same amount on each side.*

Exercise 18k

$2x + 3 - x + 5 = 3x + 4x - 6$

$$2x + 3 - x + 5 = 3x + 4x - 6$$

Simplify each side $\qquad x + 8 = 7x - 6$

Take x from both sides $\qquad 8 = 6x - 6$

Add 6 to both sides $\qquad 14 = 6x$

Divide both sides by 6 $\qquad \dfrac{14}{6} = x$

$$x = \dfrac{7}{3} = 2\dfrac{1}{3}$$

Solve the following equations:

1 $3x + 2 + 2x = 7$

2 $7 + 3x - 6 = 4$

3 $6 = 5x + 2 - 4x$

4 $9 + 4 = 3x + 4x$

5 $3x + 2x - 4x = 6$

6 $7 = 2 - 3 + 4x$

7 $5x + x - 6x + 2x = 9$

8 $5 + x + x = 1 + 4x$

9 $5x + 6 + 3x = 10$

10 $8 = 7 - 11 + 6x$

11 $7 + 2x = 12x - 7x + 2$

12 $1 + 4 - 3 + 2x = 3x$

13 $3x + 4x - x = x + 6$

14 $2 + 4x - x = x + 8$

15 $4 - x - 2 + x = x$

16 $3x + 1 + 2x = 6$

17 $4x - 2 + 6x - 4 = 64$

18 $2x + 7 - x + 3 = 6x$

19 $6 - 2x - 4 + 5x = 17$

20 $9x - 6 - x - 2 = 0$

21 $x - 3 + 7x + 9 = 10$

22 $15x + 2x - 6x - 9x = 20$

 Investigation

Meg wanted to find out Malcolm's age without asking him directly what it was. The following conversation took place.

Meg: Think of your age but don't tell me what it is

Malcolm: Right

Meg: Multiply it by 5, add 4 and take away your age.

Malcolm: Yes

Meg: Divide the result by 4 and tell me your answer.

Malcolm: 15

Meg: That means you are 14.

Malcolm: Correct. How do you know that?

However many times Meg tried this on her friends and relations she found their age by taking 1 away from the number they gave.

1 Does it always work?

2 Can you use simple algebra to prove that it always gives the correct answer?

Mixed exercises

Exercise 18l

1 Solve the equation $3x + 2 = 4$.

2 I think of a number, add 4 and the result is 10. Form an equation and solve it to find the number I thought of.

3 Solve the equation $6x + 2 = 3x + 8$.

4 Solve the equation $4x - 2 = 6$.

5 Simplify $4x - 3y + 5x + 2y$.

6 Solve the equation $4x + 2 - x = 6$.

Exercise 18m

1 Solve the equation $4x - 5 = 3$.

2 Simplify $3c - 5c + 9c$.

3 Solve the equation $3x - 2 = 4 - x$.

4 When I think of a number, double it and add three, I get 11. What number did I think of?

5 Solve the equation $x + 2x - 4 = 9$.

6 Simplify $2a + 4 - 3 + 5a - a$.

Exercise 18n

1 If $2x - 9 = 2$, find x.

2 Simplify $6p + p - 3p - 4p$.

3 Find x if $4 + 2x = 6 + x$.

4 If $3c + 2 = c + 2$, find c.

5 Simplify $3b + 4c - 5c + c$.

6 Peter had 14 marbles and lost x of them. John started with 8 marbles and gained x. The two boys then found that they each had the same number of marbles. Form an equation and find x.

Exercise 18p

1 If $9 = 3x - 3$, find x.

2 Simplify $2x + 5x - 8x$.

3 Solve the equation $3 - 2n = 5 + 3n$.

4 If $6x - 4 = 2x + 4$, find x.

5 Simplify $3a + 2d - a + 4c + 3d + c$.

6 Solve the equation $2x + 8 + 3x - 6 = 4$.

Investigation

1   Each student must bring a calendar page for the month of his or her birth.

   a   Choose any four-by-four grid of sixteen days, not including blank squares. Outline this grid and find
       the sum of the four centre numbers

       the sum of the four numbers on each diagonal.
   b   What do you notice?
   c   Compare your results with those of other members of the class.

2   Do calendars from the same month give the same or different sums when using different four-by-four squares?

3   What happens if other months are used that begin on a different day?

In this chapter you have seen that...

✔ an expression is a collection of terms without an 'is equal to' sign

✔ like terms are groups of numbers or groups of terms with the same letter

✔ you can add and subtract like terms to simplify them but you cannot add and subtract unlike terms

✔ an equation means the equality of two expressions

✔ you can do whatever you like with an equation as long as you do the same to both sides

✔ to solve an equation, aim to have the letter terms on one side of the equals sign and the number terms on the other side

✔ you can check your solution by replacing the letter by the number in each side independently; they will be equal if your answer is correct.

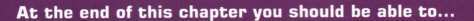

19 More algebra

Did you know?

In his treatises, René Descartes signed his name in Latin – Renatus Cartesius.
It is from this that the name Cartesian originated.

When you plot points on a graph, you will use Cartesian coordinates.

You need to know...

✔ how to distinguish between like and unlike terms

✔ how to collect like terms

✔ that if you do something to one side in any equation then you must do exactly the same to the other side

✔ how to solve simple equations in one unknown involving two or more operations

Key words

coordinates, equation, expression, formula, perimeter, product, rectangle, solve, solution

Brackets

Sometimes brackets are used to hold two quantities together. For instance, if we wish to multiply the sum of x and 3 by 4 we write $4(x+3)$. The multiplication sign is invisible just as it is in $5x$, which means $5 \times x$.

$4(x+3)$ means 'four times everything in the brackets'

so we have $4 \times x$ and 4×3, and we write $4(x+3) = 4x+12$.

Exercise 19a

Multiply out: $2(x+1)$

Multiply every term in the brackets by the number in front of the bracket.

$$2(x+1) = 2x+2$$

Multiply out the following brackets:

1	$2(x+1)$	**5**	$2(4+5x)$	**9**	$3(6+4x)$
2	$3(3x+2)$	**6**	$2(6+5a)$	**10**	$5(x+1)$
3	$5(x+6)$	**7**	$5(a+b)$	**11**	$7(2+x)$
4	$4(3x+3)$	**8**	$4(4x+3)$	**12**	$8(3+2x)$

Multiply each term in the bracket by 2.

To simplify an expression containing brackets we first multiply out the brackets and then collect like terms.

Exercise 19b

Simplify: $2+(3x+7)$

First deal with the brackets, then collect like terms.

$$2+(3x+7) \qquad \text{(This means } 2+1(3x+7))$$
$$= 2+3x+7$$
$$= 3x+9$$

Simplify:

1	$2x+4(x+1)$	**5**	$2(x+4)+3(x+5)$
2	$3+5(2x+3)$	**6**	$3x+(2x+5)$
3	$3(x+1)+4$	**7**	$4+(3x+1)$
4	$7+2(2x+5)$	**8**	$3x+2(3x+4)$

Multiply out both brackets, then collect like terms.

Equations containing brackets

If we wish to solve equations containing brackets we first multiply out the brackets and then collect like terms.

Exercise 19c

Solve: $4 + 2(x + 1) = 22$

Remember whatever you do to one side of an equation you must do to the other.

$$4 + 2(x + 1) = 22 \quad \text{First multiply out the brackets}$$
$$4 + 2x + 2 = 22$$
$$2x + 6 = 22 \quad \text{You want the } x \text{ term on its own.}$$

Take 6 from both sides $\quad 2x = 16$

Divide both sides by 2 $\quad x = 8$

Check: If $x = 8$, left-hand side $= 4 + 2(8 + 1)$
$$= 4 + 2 \times 9$$
$$= 22$$

Right-hand side $= 22$, so $x = 8$ is the solution.

Solve the following equations:

1 $\quad 6 + 3(x + 4) = 24$

2 $\quad 3x + 2 = 2(2x + 1)$

3 $\quad 5x + 3(x + 1) = 14$

4 $\quad 5(x + 1) = 20$

5 $\quad 2(x + 5) = 6(x + 1)$

6 $\quad 28 = 4(3x + 1)$

7 $\quad 4 + 2(x + 1) = 12$

8 $\quad 7x + (x + 2) = 22$

9 $\quad 8x + 3(2x + 1) = 7$

10 $\quad 4x - 2 = 1 + (2x + 3)$

11 $\quad 7x + x = 4x + (x + 1)$

12 $\quad 3(x + 2) + 4(2x + 1) = 6x + 20$

13 $\quad 6x + 4 + 5(x + 6) = 56$

14 $\quad 2 + 3(x + 8) = 4(2x + 1)$

Hindu problem-solving

The ancient Hindus were fond of doing number puzzles. A mathematician named Aryabhata who lived in India during the 6th century CE liked this kind of puzzle:

If 5 is added to a certain number, the result divided by 2, that result multiplied by 6, and then 8 subtracted from that result, the answer is 34. Find the number.

Aryabhata solved this problem using the method of inversion, i.e. he worked backwards and did the inverse, or opposite steps as he went along. Adding and subtracting are inverse steps. Dividing and multiplying are inverse steps.

Let us set out Aryabhata's problem using a diagram.

We will set out the problem on the left starting at the top.

On the right we will start at the bottom and do the inverse steps:

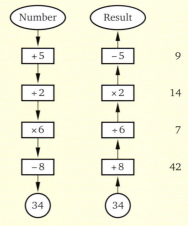

Try this method yourself on questions 1, 2 and 4 of the next exercise.

Problems to be solved by forming equations

Exercise 19d

The width of a rectangle is x cm. Its length is 4 cm more than its width. The perimeter is 48 cm. What is the width?

The width is x cm so the length is $(x+4)$ cm.

The perimeter is the distance all around the rectangle; from the diagram this is $x + (x+4) + x + (x+4)$.

You also know this is 48 cm

so $x + (x+4) + x + (x+4) = 48$

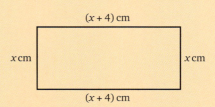

Removing the brackets and collecting like terms gives

$4x + 8 = 48$

Take 8 from each side $4x = 40$

Divide each side by 4 $x = 10$

Therefore the width is 10 cm.

A choc-ice costs x dollars and a cone costs 30 dollars more. One choc-ice and two cones together cost 540 dollars.

How much is a choc-ice?

A choc-ice costs $\$x$ and a cone costs $\$(x+30)$.

It follows that 2 cones cost $2 \times (x+30)$ dollars
so the cost of a choc-ice and two cones is $x+2(x+30)$ dollars

But you know that the total cost is $540

so
$$x+2(x+30) = 540 \quad \text{Now multiply out the bracket}$$
$$x+2x+60 = 540 \quad \text{Collect like terms}$$
$$3x+60 = 540 \quad \text{Take 60 from each side}$$
$$3x = 480 \quad \text{Divide each side by 3}$$
$$x = 160$$

Therefore a choc-ice costs $160.

Solve the following problems by forming an equation in each case.
Explain, either in words or on a diagram, what your letter stands for and always end by answering the question asked.

1 I think of a number, double it and add 14. The result is 36. What is the number?

Start by letting the number be x. Next write down what double this is and then add 14. You can now form an equation in x.

2 I think of a number and add 6. The result is equal to twice the first number. What is the first number?

3 In triangle ABC, AB = AC. The perimeter is 24 cm. Find AB.

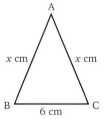

A

x cm x cm

B — 6 cm — C

4 A bun costs x dollars and a cake costs 3 dollars more than a bun. Four cakes and three buns together cost $159. How much does one bun cost?

Find the cost of 1 cake, then the cost of 4 cakes. Add on the cost of 3 buns. The total cost is given, so you can form an equation in x and solve it to find x.

5 A bus started from the terminus with x passengers. At the first stop another x passengers got on and 3 got off. At the next stop, 8 passengers got on. There were then 37 passengers. How many passengers were there on the bus to start with?

6 Buns cost x dollars each and a cake costs twice as much as a bun. I buy two buns and three cakes and pay $160 altogether. How much does one bun cost?

7 I think of a number, add 6, multiply by 2 and the result is 20. What is the number?

8

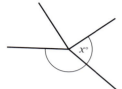

The first angle measures $x°$, the second angle (going clockwise) is twice the first, the third is 30° and the fourth is 90°. Find the first angle.

9 30 sweets are divided amongst Anne, Mary and John. Anne has x sweets, Mary has three times as many as Anne, and John has 6. How many sweets has Anne?

Formulae

A formula is an instruction for working out a quantity.

A supplier of kitchen gadgets uses this formula for working out the delivery charges.

Delivery charge = $1000 plus $200 per item.

So the delivery charge for 6 items is $1000 + 6 × $200 = $2200.

We can shorten the formula if we use letters rather than words for the variable quantities. Using D for the delivery charge and n for the number of items, we can write $D = 1000 + 200n$.

For all rectangles it is true that the area is equal to the length multiplied by the breadth, provided that the length and breadth are measured in the same units.

If we use letters for the unknown quantities (A for area, l for length, b for breadth) we can write the first sentence more briefly as a formula: $A = l × b$.

The multiplication sign is usually left out giving

$$A = lb$$

Exercise 19e

The letters in the diagrams all stand for a number of centimetres.

The perimeter of the square below is P cm. Write down a formula for P.

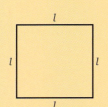

Start by writing the perimeter in terms of the letters in the diagram: this is $l+l+l+l$ (cm). As we are told that P cm is the perimeter we can write

$$P = l+l+l+l$$

Collect like terms $P = 4l$

In each of the following figures the perimeter is P cm. Write down a formula for P starting with $P =$

1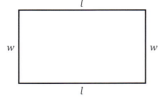

2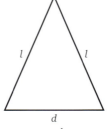

3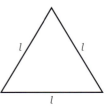

4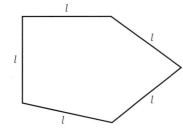

5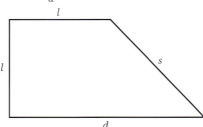

If G is the number of girls in a class and B is the number of boys, write down a formula for the total number, T, of children in the class.

Start by writing the total number in terms of the letters other than T.

$$T = G+B$$

6 I buy x kg of mangos and y kg of papayas. Write down a formula for W if W kg is the mass of fruit that I have bought.

Read each question carefully.

7 If l m is the length of a rectangle and b m is the breadth, write down a formula for P if the perimeter of the rectangle is P m.

8   I start a game with N marbles and win another M marbles. Write down a formula for the number, T, of marbles that I finish the game with.

9   I start a game with N marbles and lose L marbles. Write down a formula for the number, T, of marbles that I finish with.

10 The side of a square is l m long. Write down a formula for A if the area of the square is A m².

11 Breadfruit cost $\$n$ each. Write down a formula for N if the cost of 10 breadfruit is $\$N$.

12 Oranges cost x dollars each and I buy n of these oranges. Write down a formula for C where C dollars is the total cost of the oranges.

13 I have a piece of string which is l cm long. I cut off a piece that is d cm long. Write down a formula for L if the length of string that is left is L cm.

14 A rectangle is $2l$ m long and l m wide. Write down a formula for P where P m is the perimeter of the rectangle.

15 Write down a formula for A where A m² is the area of the rectangle described in question 14.

16 I had a bag of sweets with S sweets in it; I then ate T of them. Write down a formula for the number, N, of sweets left in the bag.

17 A lorry has mass T tonnes when empty. Steel girders with a total mass of S tonnes are then loaded on to the lorry. Write down a formula for W where W tonnes is the mass of the loaded lorry.

18 I started the term with a new packet of N felt-tipped pens. During the term I lost L of them and R of them ran dry. Write down a formula for the number, S, that I had at the end of the term.

19 A truck travels p km in one direction and then it comes back q km in the opposite direction. If it is then r km from its starting point, write down a formula for r.

20 One box of tinned fruit has mass K kg. The mass of n such boxes is W kg. Write down a formula for W.

21 Two points have the same y-coordinate. The x-coordinate of one point is a and the x-coordinate of the other point is b. If d is the distance between the two points, write down a formula for d given that a is less than b.

Make a sketch to illustrate this problem.

22 A letter costs x dollars to post. The cost of posting 20 such letters is $\$q$. Write down a formula for q.

23 One grapefruit weighs y g. The weight of n such grapefruit is L kg. Write down a formula for L.

Look carefully at the units.

24 A rectangle is l m long and b cm wide. The area is A cm². Write down a formula for A.

25 On my way to work this morning the bus I was travelling on broke down. I spend t hours on the bus and s minutes walking. Write down a formula for T if the total time that my journey took was T hours.

Substituting numerical values into a formula

The formula for the area of a rectangle is $A = lb$.

If a rectangle is 3 cm long and 2 cm wide, we can substitute the number 3 for l and the number 2 for b to give $A = 3 \times 2 = 6$.

So the area of that rectangle is 6 cm².

When you substitute numerical values into a formula you may have a mixture of operations, i.e. ( ), ×, ÷, +, −, to perform. Remember the order from the capital letters of 'Bless My Dear Aunt Sally'.

Exercise 19f

If $v = u + at$, find v when $u = 2$, $a = \frac{1}{2}$ and $t = 4$.
$$v = u + at$$
When $u = 2$, $a = \frac{1}{2}$, $t = 4$, $v = 2 + \frac{1}{2} \times 4$ (Do multiplication first)
$$= 2 + 2$$
$$= 4$$

1 If $N = T + G$, find N when $T = 4$ and $G = 6$.

2 If $T = np$, find T when $n = 20$ and $p = 5$.

3 If $P = 2(l + b)$, find P when $l = 6$ and $b = 9$.

4 If $L = x - y$, find L when $x = 8$ and $y = 6$.

5 If $N = 4(l - s)$, find N when $l = 7$ and $s = 2$.

6 If $S = n(a + b)$, find S when $n = 20$, $a = 2$ and $b = 8$.

7 If $V = lbw$, find V when $l = 4$, $b = 3$ and $w = 2$.

8 If $A = \frac{PRT}{100}$, find A when $P = 100$, $R = 3$ and $T = 5$.

9 If $w = u(v - t)$, find w when $u = 5$, $v = 7$ and $t = 2$.

10 If $s = \frac{1}{2}(a + b + c)$, find s when $a = 5$, $b = 7$ and $c = 3$.

Given that $2S = d(a + l)$, find a when $S = 20$, $d = 2$ and $l = 16$

$$2S = d(a + l)$$

Substituting $S = 20$, $d = 2$, $l = 16$, gives

$$40 = 2(a + 16)$$

(We can now solve this equation for a.)

Multiply out the brackets $40 = 2a + 32$

Take 32 from each side $8 = 2a$

Divide by 2 $4 = a$ or $a = 4$

---

11 Given that $N = G + B$, find B when $N = 40$ and $G = 25$.

12 If $R = t \times c$, find t when $R = 10$ and $c = 20$.

13 Given that $d = st$, find t when $d = 50$ and $s = 15$.

14 If $N = 2(p + q)$, find q when $N = 24$ and $p = 5$.

15 Given that $L = P(2 - a)$, find a when $L = 10$ and $P = 40$.

16 Given that $s = 3(a - b)$, find b when $s = 15$ and $a = 24$.

17 Given that $v = u + at$, find u when $v = 32$, $a = 8$ and $t = 4$.

Problems

Exercise 19g

A rectangle is $3l$ cm long and l cm wide. If the area of the rectangle is A cm², write down a formula for A.

Use your formula to find the area of this rectangle if it is 5 cm wide.

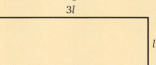

Area = length × width

$\therefore$ $A = 3l \times l$

When $l = 5$, $A = 3 \times 5 \times 5 = 75$

$\therefore$ Area = 75 cm²

1   Oranges cost $n each. If the cost of a box of 50 of these oranges is $C, write down a formula for C. Use your formula to find the cost of a box of oranges if each orange costs $12.

2   Lemons cost $n each. The cost of a box of 50 lemons is $L. Write down a formula for L. Use your formula to find the cost of a box of these lemons when they cost $10 each.

3   A rectangle is a cm long and b cm wide. Write down a formula for P if P cm is the perimeter of the rectangle. Use your formula to find the perimeter of a rectangle measuring 20 cm by 15 cm.

4   The length of a rectangle is twice its width. If the rectangle is x cm wide, write down a formula for P if its perimeter is P cm. Use your formula to find the width of a rectangle that has a perimeter of 36 cm.

5   A roll of paper is L m long. N pieces each of length r m are cut off the roll. If the length of paper left is P m, write down a formula for P. Use your formula to find the length of paper left from a roll that was 20 m long after 10 pieces, each of length 1.5 m, are cut off.

6   Tins of baked beans weigh a g each. N of these tins are packed into a box. The empty box weighs p g. Write down a formula for W where W g is the weight of the full box. Use your formula to find the number of tins that are in a full box if the full box weighs 10 kg, the empty box weighs 1 kg and each tin weighs 200 g.

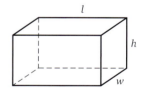

Be careful with the units.

7   The rectangular box in the diagram is l cm long, w cm wide and h cm high. Write down a formula for A if A cm² is the total surface area of the box (i.e. the area of all six faces). Use your formula to find the surface area of a rectangular box measuring 50 cm by 30 cm by 20 cm.

8   A person whose weight on Earth is W finds his weight on certain planets from these formulae.
    a   Weight on Venus 0.85 W.
    b   Weight on Mars 0.38 W.
    c   Weight on Jupiter 2.64 W.
    Calculate your weight on each of the above planets.

Mixed exercises

Exercise 19h

1   Solve the equation $2x - 3 = 7$.
2   Simplify $4(x + 3) + 1$.

3 I think of a number, double it, add 6 and the result is 32.
Find the number.

4 Solve the equation $3x + 1 = 2x + 3$.

5 If $P = a - b$, find the value of P when $a = 5$ and $b = 2$.

6 If $R = N - D$, find the value of N when $R = 4$ and $D = 7$.

7 Write down a formula for l if P is the perimeter of this figure:

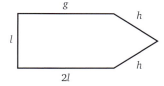

Exercise 19i

1 Solve the equation $5x + 3 = 6 - x$.

2 Simplify $4x + 3(2x + 5)$.

3 I think of a number, double it, subtract 10 and my answer is 2 more
than the number I first thought of. What was the number?

4 Solve the equation $4(x + 3) = 3(2x + 4)$.

5 Solve the equation $2x + (x - 3) = 0$.

6 Find the value of N given that $N = 5s - t$ when $s = 10$ and $t = 20$.

7 A rectangle is twice as long as it is wide. It is a cm wide. Write down
a formula for P where P cm is the perimeter of the rectangle.

Exercise 19j

1 Solve the equation $4 + 2x = 2(3 - x)$.

2 Andrew has 6 sweets, Mary has x sweets and Jim has twice as many
as Andrew. Together they have four times as many as Mary has.
Form an equation and find how many sweets Mary has.

3 Simplify $3 + 2(x + 2)$.

4 Solve the equation $3x + 7 = x + 9$.

5 A cell phone company charges $2000 a month together with
$20 per minute for phone calls and $10 for each text.
  a Find a formula for N, the monthly cost of making calls lasting a
  total of m minutes and sending t texts.
  b Find the bill for June when David made calls totalling 200 minutes
  and sent 500 texts.

6   When shopping, Mrs Jones spent $x in the first shop, the same amount in the second shop, $2 in the third and $8 in the last. The total amount she spent was $18. Form an equation. How much did she spend in the first shop?

7   Simplify $4+3(x+2)$.

8   There is no solution of the equation $4x+3+2x=6x$. Find the reason by trying to solve the equation.

(?) Puzzle

May bought a certain number of $190 tickets and a certain number of $280 tickets.

Altogether she paid $3190 for these tickets. How many of each did she buy?

(!) Investigation

P is any point on a diagonal of a $6\,\text{cm} \times 4\,\text{cm}$ rectangle.

Investigate the relationship between the area of rectangle A and the area of rectangle B.

Which of these is true?
area A > area B
area A = area B
area A < area B

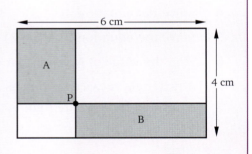

In this chapter you have seen that...

✔ you can simplify algebraic expressions involving brackets by multiplying out the brackets

✔ you can solve problems by forming an equation

✔ you can use a formula to work out the value of a quantity

✔ you can use letters for variable quantities.

Did you know?

Florence Nightingale gathered statistics during the Crimean War (1853–1856) and put them to good use. When she arrived at Scutari she found that 42% of the soldiers in the hospital died, mainly from disease, not war wounds. With a team of nurses she reduced this figure to 2%. She went on to found the first professional nursing service.

You need to know...

✔ how to use a protractor to draw and measure angles
✔ how to find fractions of a quantity

Key words

bar chart, circle, data, frequency, frequency column, frequency table, horizontal, pictograph, pie chart, protractor, radius, range, tally

Frequency tables

The branch of mathematics called statistics is used for dealing with large collections of information in the form of numbers. The number of items of information can run into thousands when the incomes of everyone in Jamaica are being considered, for instance, but to learn the methods we start with smaller collections.

If we collect the heights in centimetres of 72 children in Grade 7 we are faced with a disorganised set of numbers:

147	146	151	137	149	159	142	150	151
138	139	155	151	152	145	139	135	153
139	151	145	162	152	138	142	140	155
146	165	155	149	162	145	152	148	152
132	152	142	152	152	143	145	157	152
148	145	154	145	149	155	137	144	140
139	145	151	152	152	140	160	155	151
136	151	149	151	156	142	134	156	166

To make sense of these numbers we must put them in order. One way of doing this is to form a *frequency table*. We do not always wish to write down every number so we group them, as in the table below. Work down the columns, making a tally mark, |, in the tally column opposite the appropriate group. (Do *not* go through the columns looking for numbers that fit into the first group and the second group and so on.) Count up the tally marks and write the total in the frequency column. Check by adding up the numbers in the frequency column.

(Arrange the tally marks in fives either by leaving a gap between blocks as is done in this table or by crossing four tally marks with the fifth thus:
卌 卌 ||)

Height in cm (correct to nearest cm)	Tally	Frequency
131–135	|||	3
136–140	||||| ||||| ||	12
141–145	||||| ||||| |||	13
146–150	||||| |||||	10
151–155	||||| ||||| ||||| ||||| |||||	25
156–160	|||||		6
161–165	|||	3
	Total	72

We can see now that there are a few children with small or large heights and that the greatest number have heights in the 151–155 cm range.

Exercise 20a

Draw up tables like the one on the previous page, using the groups suggested.

1   The following numbers are the heights in centimetres of the same children as those on page 288, when they had reached Grade 9:

154	166	153	166	149	154	153	160	165
164	156	166	156	166	161	155	164	164
156	159	161	150	163	163	154	157	159
150	146	157	168	167	154	166	150	157
154	162	164	152	154	153	163	157	163
161	168	150	152	163	164	157	159	160
164	158	158	165	167	170	156	164	164
152	155	163	164	157	166	161	148	168

Use groups 146–150 cm, 151–155 cm. 156–160 cm, 161–165 cm, 166–170 cm.

2   The following numbers are the masses in kilograms (to the nearest kg) of the same 72 children in Grade 9:

41	50	54	52	65	54	48	50	43
48	58	46	43	50	48	47	44	48
43	44	47	45	57	54	42	52	49
47	40	53	41	41	49	44	59	43
35	51	44	44	49	45	62	46	51
55	54	54	41	43	70	40	44	59
45	43	45	37	51	39	55	53	45
61	44	57	39	51	44	48	44	51

Use groups 35–39 kg, 40–44 kg, 45–49 kg, 50–54 kg, 55–59 kg, 60–64 kg, 65–70 kg, 70–74 kg.

3   The following are the marks of 82 pupils in a mathematics examination:

78	41	56	66	76	65	50	37	45	40
87	38	49	82	41	79	66	95	19	38
31	75	54	49	65	53	69	63	67	91
62	34	79	84	71	85	42	59	74	56
56	50	53	68	61	54	25	64	84	80
48	64	72	53	44	55	35	63	36	81
70	73	47	63	42	57	51	63	52	45
38	62	64	47	62	48	28	60	61	58
57	39								

Use groups 11–20, 21–30, 31–40, and so on.

Bar charts

The information collected can be illustrated in various ways and one of the most common is the bar chart.

Here is a bar chart to show the heights of 72 children in the first year (using the table on page 288).

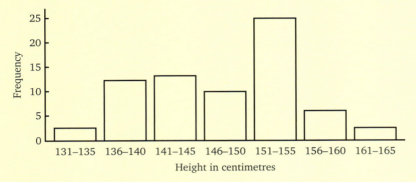

The bars must be all the same width but they do not have to touch.
The spaces between the bars must all be the same.

Notice that the groups are arranged along the base line and the frequencies are marked on the vertical axis.

We can see at a glance that the greatest number of children have heights between 151–155 cm.

A group of people were asked to select their favourite colour from a card showing 6 colours and the following results were recorded:

Colour	Rose pink	Sky blue	Golden yellow	Violet	Lime green	Tomato red
Number of people (frequency)	6	8	8	2	1	10

The bar chart below shows people's favourite colours:

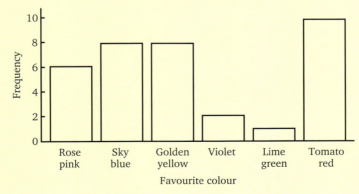

Notice that the bars are all the same width.

Exercise 20b

In questions **1** to **5** draw bar charts to show the information given in the frequency tables. Mark the frequency on the vertical axis and label the bars below the horizontal axis.

1 Types of vehicles moving along a busy road during one hour:

Vehicle	Cars	Vans	Lorries	Motorcycles	Bicycles
Frequency	62	11	15	10	2

2 Thirty pupils were asked to state their favourite subject chosen from their school timetable:

Subject	English	Mathematics	French	PE	History	Geography
Frequency	5	7	4	3	7	4

3–5 Use the information from the frequency tables in exercise **20a** (page 289), numbers **1** to **3**.

6 The number of tourist arrivals by country, to a certain Caribbean island, in the first six months of 2006 (to the nearest thousand) were:

Country	USA	Canada	UK	Caricom	Other
No. of tourists	80 000	46 000	18 000	31 000	15 000

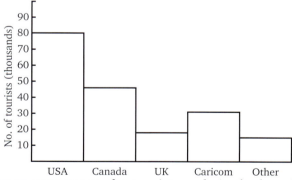

In an attempt to save space, the tourists scale in the bar chart shown above was started at 15 000 instead of at 0.

a Redraw the bar chart with the tourist scale from 0 to 90 000 (suggested scale 1 cm to 15 000).

b Compare the two bar charts. The impression given by one of them is misleading. Why?

Bar charts can be used to represent information other than frequencies and can appear in different forms. The bars are usually vertical but occasionally they are horizontal.

1   The average price per tonne of sugar earned by a Caribbean country
    in the ten-year period 1987–1996:

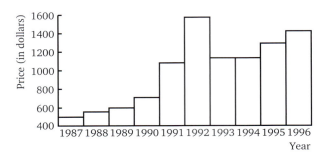

a   In which year was the price lowest? Highest?
b   In which year was the price increase from the previous year the
    greatest?
c   In which years was the price of sugar above $1400 per tonne?
d   In which two years did the price remain the same?

2   The tourist arrivals (in thousands) to Antigua and St Lucia for the
    period 1987–1996:

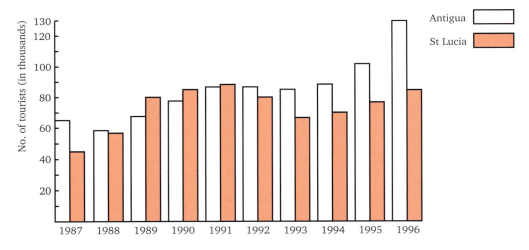

a   Did more tourists visit St Lucia or Antigua during the period
    1987–1996?
b   In which year did St Lucia have fewest tourists?
c   In which years did St Lucia have better tourist seasons than Antigua?

3 Cost of defence, health, education and housing over a three-year period:

No numbers are given, but we can get a good idea about the relative costs.

a In which year was the most money spent?

b On what was the least money spent?

c In which period did education cost most?

d In which period did health cost least?

e In which period was the least money spent?

? Puzzle

A rough guide to the distance to keep behind another car on the road:

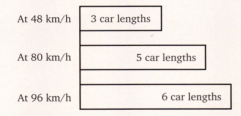

Can you puzzle out what rule has been used to decide on the distances?

Why is the guide only 'rough'? What other factors should be taken into account?

Exercise 20d

1 a Draw a chart to show the number of thousands of tonnes of alumina produced in Guyana during the 11 years 1986–1996.

Year	1986	1987	1988	1989	1990	1991	1992	1993	1994	1995	1996
Thousands of tonnes	312	305	257	234	311	294	165	173	236	160	212

b In which year was the least alumina produced?

c In which year was most alumina produced?

2 **a** Draw a bar chart to show the birth rates per 1000 population in ten Caribbean countries in 1996.

Country	Antigua	Barba-dos	Domi-nica	Grenada	Guade-loupe	Jamaica	St Kitts	Trinidad	St Vin-cent
Birth rate per 1000	24	21	39	30	28	34	25	27	36

  b Which country has the highest birth rate?

  c Which country has the lowest birth rate?

Pie charts

A pie chart is used to represent information when some quantity is shared out and divided into different categories.

Here is a pie chart to show the proportions, within a group, of people with eyes of certain colours.

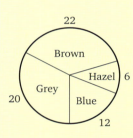

The size of the 'pie slice' represents the size of the group. We can see without looking at the numbers that there are about the same number of people with brown eyes as with grey eyes and that there are about twice as many with grey eyes as with blue. The size of the pie slice is given by the size of the angle at the centre, so to draw a pie chart we need to calculate the sizes of the angles.

The number of people is 60.

As there are 12 blue-eyed people, they form $\frac{12}{60}$ of the whole group and are therefore represented by that fraction of the circle.

Blue: $\frac{12}{60} \times \frac{360°}{1} \times 1 = 72°$ Grey: $\frac{20}{60} \times \frac{360°}{1} = 120°$

Hazel: $\frac{6}{60} \times \frac{360°}{1} = 36°$ Brown: $\frac{22}{60} \times \frac{360°}{1} = 132°$

Total = 360°

Now draw a circle of radius about 5 cm (or whatever is suitable). Draw one radius as shown and complete the diagram using a protractor, turning your page into the easiest position for drawing each new angle.

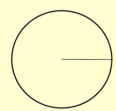

Label each 'slice'.

Exercise 20e

Draw pie charts to represent the following information, first working out
the angles.

1 A box of 60 coloured balloons contains the following
numbers of balloons of each colour:

Colour	Red	Yellow	Green	Blue	White
Number of balloons	16	22	10	7	5

Find the number of balloons in each category as a fraction of the total number of balloons. Then find this fraction of 360°.

2 Ninety people were asked how they travelled to work and the
following information was recorded:

Transport	Car	Bus	Walk	Motorcycle	Bicycle
Number of people	32	38	12	6	2

3 On a cornflakes packet the composition of 120 g of cornflakes is given
in grams as follows:

Protein	Fat	Carbohydrate	Other ingredients
101	1	10	8

4 Of 90 cars passing a survey point it was recorded that 21 had
two doors, 51 had four doors, 12 had three (two side doors and a
hatchback) and 6 had five doors.

5 A large flower arrangement contained 18 dark red roses, 6 pale pink
roses, 10 white roses and 11 deep pink roses.

6 Use the information given in exercise **20b**, number **2**, page 291.

7 The children in a class were asked what pets they owned and the
following information was recorded:

Animal	Dog	Cat	Bird	Rabbit	Fish
Frequency	8	10	3	6	3

Sometimes the total number involved is not as convenient as in the previous
problems. We may have to find an angle correct to the nearest degree.
If there had been 54 people whose eye colours were recorded we might have
had the following information:

Eye colour	Blue	Grey	Hazel	Brown
Frequency	10	19	5	20

Total = 54

Angles

Blue: $\frac{10}{54} \times \frac{360°}{1} = \frac{200°}{3}$

$= 66\frac{2}{3}° = 67°$ (to the nearest degree)

Grey: $\frac{19}{54} \times \frac{360°}{1} = \frac{380°}{3}$

$= 126\frac{2}{3}° = 127°$ (to the nearest degree)

Hazel: $\frac{5}{54} \times \frac{360°}{1} = \frac{100°}{3}$

$= 33\frac{1}{3}° = 33°$ (to the nearest degree)

Brown: $\frac{20}{54} \times \frac{360°}{1} = \frac{400°}{3}$

$= 133\frac{1}{3}° = 133°$ (to the nearest degree)

Total $= 360°$

Draw pie charts to represent the following information, working out the angles first and, where necessary, giving the angles correct to the nearest degree.

8 300 people were asked whether they lived in a flat, a house, a room, an apartment or in some other type of accommodation and the following information was recorded:

Type of accommodation	Flat	House	Room	Apartment	Other
Frequency	90	150	33	15	12

9 In a street in which 80 people live the numbers in various age groups are as follows:

Age group (years)	0–15	16–21	22–34	35–49	50–64	65 and over
Number of people	16	3	19	21	12	9

10 Use the information given in exercise **20b**, number **1**, page 291.

11 Use the information on people's choice of eye colour given on page 290.

Interpreting pie charts

Exercise 20f

1 This pie chart shows the uses of personal computers in 2010:

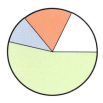

Key:
- Home and hobby
- Educational
- Scientific
- Business and professional

a For which purpose were computers used most?

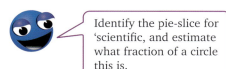
What does the biggest pie-slice represent?

b Estimate the fraction of the total sales used for

Identify the pie-slice for 'scientific, and estimate what fraction of a circle this is.

  i scientific purposes
  ii home and hobbies.

2 The pie chart below shows how fuel is used for different purposes in the average house.

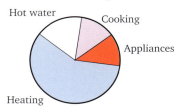

Hot water, Cooking, Appliances, Heating

Is the angle of the slice for 'cooking' bigger or smaller than the angle of the slice for 'hot water' and by roughly how much?

a For which purpose is most fuel used?

b How does the amount used for cooking compare with the amount used for hot water?

3 The pie chart shows the age distribution of a population in years in 2009:

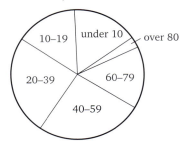

10–19, under 10, over 80, 20–39, 60–79, 40–59

a Estimate the size of the fraction of the population in the age groups

Remember that a whole turn is 360°. What fraction of a whole turn represents each age group?

  i under 10 years
  ii 20–39 years.

b State which groups are of roughly the same size.

Pictographs

To attract attention, *pictographs* are often used on posters and in newspapers and magazines. The best pictographs give the numerical information as well; the worst give the wrong impression.

Exercise 20g

1   Road deaths in the past 4 years at an accident black spot:

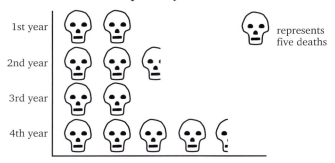

a   Give an estimate of the number of deaths in each year.

b   What message is the poster trying to convey?

c   How effective do you think it is?

2   The most popular subject among Grade 7 pupils:

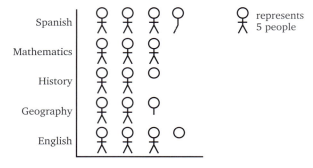

a   Which is the most popular subject?

b   How many pupils chose each subject and how many were asked altogether?

c   Is this a good way of presenting the information?

3 This is a bar chart in an advertisement showing the consumption of Fizz lemonade:

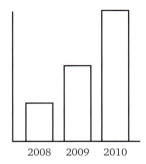

a What does this show about the consumption of lemonade?

It was decided to change from a bar chart to a pictograph for the next advertisement:

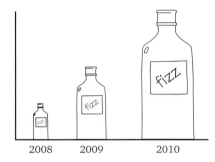

b This looks impressive but it could be misleading. Why?

Drawing pictographs

Make sure when using drawings that each takes up the same amount of space and is simple and clear.

Exercise 20h

1 Eighty-five people were asked how they travelled to work and the following information was recorded:

Transport	Car	Bus	Walk	Bicycle
Number of people	30	40	10	5

Draw a pictograph using one drawing to 5 people.

2 Thirty pupils in a class were asked what they were writing with. The following information was recorded:

Writing implement	Black pen	Blue pen	Pencil
Frequency	12	9	9

3   Some children were asked what pets they owned:

Pet	Dog	Cat	Bird	Small animal	Fish
Frequency	9	7	6	10	2

Use one drawing to one pet. Make the symbols simple.

The symbol for fish could be

! Investigation

Collect the information; where it is necessary, decide on the groups it can be divided into and record the information in a frequency table as on page 288. Decide whether a bar chart, a pie chart or a pictograph would be most suitable for presenting the information.

Suggestions for Class Projects

1   Heights of children in the class.

2   Weights of children in the class.

3   Handspan. Stretch your hand out as wide as it will go on a piece of paper and mark the positions of the end of the thumb and of the little finger. Measure the distance between these points to the nearest centimetre.

4   Times of journeys to school in minutes.

5   Times of arrival at school.

6   Find out the size of the family of each child in your class. Find out the numbers of boys and girls in each family. Compare the ratio of boys : girls to the national figures for this ratio.

7   Pets owned.

8   Pets you would *like* to own, but decide on categories first before collecting the information.

9   Birthday months.

10 Number of houses in the street where a pupil lives. Decide what to record if houses are isolated or the pupil lives in an apartment.

11 Colours of cars seen passing during, say, 20 minutes.

12 Number of people in cars travelling at a given time of day, say on the way to school.

Suggestions for Individual Projects

13 Toss one die 120 times and record the scores.

14 Toss two dice 120 times and record the combined score each time.

15 Choose a page of a book of plain text and record the occurrence of the different letters of the alphabet.

16 Choose a page of text in a different language and repeat number 15. Compare the two sets of results.

17 Choose pages of text from a book and record the lengths of, say, 60 sentences.

18 Choose a page of text and record the number of letters used in each word. Decide beforehand what to do about words with hyphens.

19 What is the most common first name for boys in the school? For girls? Consider using one class, a whole class level or the entire school to find out.

 Investigation

From time to time the government of Jamaica take a census.

What is a census? How often is it taken? Who organises it? How are the data collected?

What is the information used for?

In this chapter you have seen that...

✔ large quantities of information can be made sense of by grouping it and putting it into a frequency table. The frequency of a group is the number of items in that group

✔ frequency tables can be represented by bar charts or pie charts or pictograms

✔ the heights of the bars correspond to the frequencies of the groups

✔ the slices on a pie chart represent the fractions that the groups are of the total.

21 Probability

At the end of this chapter you should be able to...

1 Write down the set of outcomes of an experiment.

2 State the number of possible outcomes of an experiment.

3 Calculate the probability of an event A happening as

$$P(A) = \frac{\text{the number of ways in which } A \text{ can occur}}{\text{the total number of equally likely outcomes.}}$$

4 Calculate the probability that event A does not happen as

$1 - P(A \text{ does happen})$

5 Estimate the number of times an event might occur.

6 Perform experiments, collect data and hence calculate required probabilities.

You need to know...

✔ how to simplify a fraction

✔ how to add and subtract fractions

✔ the cards in an ordinary pack of playing cards

Key words

approximation, biased, certainty, chance, equally likely events, equally likely outcomes, event, expectation, experiment, fair, impossibility, integer, odds, outcome, prime numbers, probability, random, unbiased

X	0	0
	X	
0		X

Do you know how to play Tic-Tac-Toe in such a way as never to lose?
If you do, you are using a topic in mathematics called Game Theory.
We told you that maths was fun!

A game is won either by chance or by using a strategy.

Probability and Statistics as mathematical sciences arose because game players wanted to use game odds to best advantage.

After studying mathematics carefully, mathematicians defined numbers called *probabilities* that helped them to explain how likely some 'outcome' or 'event' was to happen or fail to happen.

Outcomes of experiments

If you throw an ordinary die there are six possible scores that you can get. These are 1, 2, 3, 4, 5, or 6. The act of throwing the die is called an *experiment*. The score that you get is called an *outcome* or an *event*. The set {1, 2, 3, 4, 5, 6} is called the *set of all possible outcomes*.

Exercise 21a

How many possible outcomes are there for the following experiments? Write down the set of all possible outcomes in each case.

1 Tossing a 10 c coin. (Assume that it lands flat.)

2 Taking one disc from a bag containing 1 red, 1 blue and 1 yellow disc.

3 Choosing one number from this list: 1, 2, 3, 4, 5, 6, 7, 8, 9, 10.

4 Taking one crayon from a box containing 1 red, 1 yellow, 1 blue, 1 brown, 1 black and 1 green crayon.

5 Taking one item from a bag containing 1 packet of chewing gum, 1 packet of boiled sweets and 1 bar of chocolate.

6 Taking one coin from a bag containing one 1 c coin, one 10 c coin, one 25 c coin and one $1 coin.

7 Choosing one card from part of a pack of ordinary playing cards containing just the suit of clubs.

8 Choosing one letter from the vowels of the alphabet.

9 Choosing one number from the first 5 prime numbers.

10 Choosing an even number from the first 20 whole numbers.

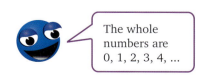

The whole numbers are 0, 1, 2, 3, 4, ...

Probability

If you throw an ordinary die, what are the chances of getting a four? If you throw it fairly, it is reasonable to assume that you are as likely to throw any one score as any other, i.e. all outcomes are equally likely. As throwing a four is only 1 of the 6 equally likely outcomes you have a 1 in 6 chance of throwing a four.

'Odds' is another word in everyday language that is used to describe chances.

In mathematical language we use the word 'probability' to describe chances. We say that the probability of throwing a four is $\frac{1}{6}$.

This can be written more briefly as

$$P(\text{throwing a four}) = \frac{1}{6}$$

We will now define exactly what we mean by 'the probability that something happens'.

If A stands for a particular event, the probability of A happening is written $P(A)$ where

$$P(A) = \frac{\text{the number of ways in which A can occur}}{\text{the } total \text{ number of equally likely outcomes}}$$

We can use this definition to work out, for example, the probability that if one card is drawn at random from a full pack of ordinary playing cards, it is the ace of spades.

(The phrase 'at random' means that any one card is as likely to be picked as any other.)

There are 52 cards in a full pack, so there are 52 equally likely outcomes.

There is only one ace of spades, so there is only one way of drawing that card,

i.e. $P(\text{ace of spades}) = \frac{1}{52}$

Exercise 21b

In the following questions, assume that all possible outcomes are equally likely.

1   One letter is chosen at random from the letters in the word SALE. What is the probability that it is A?

2 What is the probability that a red pencil is chosen from a box containing 10 different coloured pencils, one of which is red?

3 What is the probability of choosing a prime number from the numbers 6, 7, 8, 9, 10?

4 What is the probability of picking the most expensive car from a range of six new cars in a showroom?

5 What is the probability of choosing an integer that is exactly divisible by 5 from the set {6, 7, 8, 9, 10, 11, 12}?

6 Two hundred tickets are sold in a raffle. If you have bought one ticket, what is the probability that you will win first prize?

7 One card is chosen at random from a pack of 52 ordinary playing cards. What is the probability that it is the ace of hearts?

8 What is the probability of choosing the colour blue from the colours of the rainbow?

9 A whole number is chosen from the first 15 whole numbers. What is the probability that it is exactly divisible both by 3 and by 4?

 Puzzle

Charlie keeps his socks in a drawer. They are all either brown or grey. One night, in the dark, due to an electricity cut, he pulls out some socks to put on. What is the smallest number of socks he must pull out to be certain of having a pair of the same colour?

Experiments where an event can happen more than once

If a card is picked at random from an ordinary pack of 52 playing cards, what is the probability that it is a five?

There are 4 fives in the pack, the five of spades, the five of hearts, the five of diamonds and the five of clubs.

That is, there are 4 ways in which a five can be picked.

Altogether there are 52 cards that are equally likely to be picked,

therefore $P(\text{picking a five}) = \frac{4}{52} = \frac{1}{13}$

Now consider a bag containing 3 white discs and 2 black discs.

If one disc is taken from the bag it can be black or white. But these are not equally likely events: there are three ways of choosing a white disc and two ways of choosing a black disc, so

$$P(\text{choosing a white disc}) = \frac{3}{5}$$

and $$P(\text{choosing a black disc}) = \frac{2}{5}$$

Exercise 21c

A letter is chosen at random from the letters of the word DIFFICULT. How many ways are there of choosing the letter I? What is the probability that the letter I will be chosen?

There are 2 ways of choosing the letter I and there are 9 letters in DIFFICULT.

$$P(\text{choosing I}) = \frac{2}{9}$$

1. How many ways are there of choosing an even number from the first 10 whole numbers?

2. A prime number is picked at random from the set {4, 5, 6, 7, 8, 9, 10, 11}. How many ways are there of doing this?

3. A card is taken at random from an ordinary pack of 52 playing cards. How many ways are there of taking a black card?

4. An ordinary six-sided die is thrown. How many ways are there of getting a score that is greater than 4?

5. A lucky dip contains 50 boxes, only 10 of which contain a prize, the rest being empty. How many ways are there of choosing a box that contains a prize?

6. A number is chosen at random from the first 10 integers. What is the probability that it is
   a. an even number
   b. an odd number
   c. a prime number
   d. exactly divisible by 3?

7. One card is drawn at random from an ordinary pack of 52 playing cards. What is the probability that it is
   a. an ace
   b. a red card
   c. a heart
   d. a picture card (include the aces)?

8   One letter is chosen at random from the word DIFFICULT.
     What is the probability that it is
     a   the letter F
     b   the letter D
     c   a vowel
     d   one of the first five letters of the alphabet?

9   An ordinary unbiased six-sided die is thrown.
     What is the probability that the score is
     a   greater than 3
     b   at least 5
     c   less than 3?

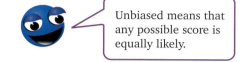

Unbiased means that any possible score is equally likely.

10   A book of 150 pages has a picture on each of 20 pages. If one page is
     chosen at random, what is the probability that it has a picture on it?

11   One counter is picked at random from a bag containing 15 red
     counters, 5 white counters and 5 yellow counters. What is the
     probability that the counter removed is
     a   red   **b**   yellow   **c**   not red?

12   If you bought 10 raffle tickets and a total of 400 were sold, what is
     the probability that you win first prize?

13   A roulette wheel is spun. What is the probability that when it stops it
     will be pointing to
     a   an even number   **b**   an odd number
     c   a number less than 10 excluding zero?

     (The numbers on a roulette wheel go from 0 to 36, and zero is
     counted neither an even number nor an odd number.)

14   One letter is chosen at random from the letters of the alphabet.
     What is the probability that it is a consonant?

15   A number is chosen at random from the set of two-digit numbers
     (i.e. the numbers from 10 to 99). What is the probability that it is
     exactly divisible both by 3 and by 4?

16   A bag of sweets contains 4 caramels, 3 fruit centres and 5 mints.
     If one sweet is taken out, what is the probability that it is
     a   a mint   **b**   a caramel   **c**   not a fruit centre?

Certainty and impossibility

Consider a bag that contains 5 red discs only. If one disc is removed it is absolutely certain that it will be red. It is impossible to take a blue disc from that bag.

$$P(\text{disc is red}) = \frac{5}{5} = 1$$

$$P(\text{disc is blue}) = \frac{0}{5} = 0$$

In all cases

$$P(\text{an event that is certain}) = 1$$

$$P(\text{an event that is impossible}) = 0$$

All events fall somewhere between the two, so

$$0 \leqslant P(\text{that an event happens}) \leqslant 1$$

Exercise 21d

Discuss the probability that the following events will happen. Try to class them as certain, impossible or somewhere in between.

1   You will swim the Atlantic Ocean.

2   You will weigh 80 kg.

3   You will be late home from school at least once this term.

4   You will grow to a height of 2 m.

5   The sun will not rise tomorrow.

6   You will run a mile in $3\frac{1}{2}$ minutes.

7   You will have a drink sometime today.

8   Jamaica will win next year's Shell Shield.

9   A card chosen from an ordinary pack of playing cards is either red or black.

10   A coin that is tossed lands on its edge.

11   Give some examples of events that are likely or unlikely to happen. For example: you will own a car; your home will burn down.

Probability that an event does not happen

If one card is drawn at random from an ordinary pack of playing cards, the probability that it is a club is given by

$$P(\text{a club}) = \frac{13}{52} = \frac{1}{4}$$

Now there are 39 cards that are not clubs so the probability that the card is not a club is given by

$$P(\text{not a club}) = \frac{39}{52} = \frac{3}{4}$$

i.e.

$$P(\text{not a club}) + P(\text{a club}) = \frac{3}{4} + \frac{1}{4} = 1$$

Hence

$$P(\text{not a club}) = 1 - P(\text{a club})$$

This relationship is true in any situation because

$$\begin{pmatrix} \text{The number of ways} \\ \text{in which an event, A,} \\ \text{can } not \text{ happen} \end{pmatrix} = \begin{pmatrix} \text{The total number of} \\ \text{possible outcomes} \end{pmatrix} - \begin{pmatrix} \text{The number of} \\ \text{ways in which A} \\ \text{can happen} \end{pmatrix}$$

i.e.

$$P(\text{A does not happen}) = 1 - P(\text{A does happen})$$

'A does not happen' is shortened to $\overline{A}$, where $\overline{A}$ is read as 'not A'.

$\overline{A}$ is sometimes written as A'.

Therefore

$$P(A') = 1 - P(A)$$

Exercise 21e

A letter is chosen at random from the letters of the word PROBABILITY. What is the probability that it is not B?

Method 1:   There are 11 letters and 2 of them are Bs

$$\therefore \quad P(\text{letter is B}) = \frac{2}{11}$$

Hence

$$P(\text{letter is not B}) = 1 - \frac{2}{11}$$

$$= \frac{9}{11}$$

Method 2:   There are 11 letters and 9 of them are not Bs

$$\therefore \quad P(\text{letter is not B}) = \frac{9}{11}$$

1   A number is chosen at random from the first 20 whole numbers. What is the probability that it is not a prime number?

2   A card is drawn at random from an ordinary pack of playing cards. What is the probability that it is not a two?

3   One letter is chosen at random from the letters of the alphabet. What is the probability that it is not a vowel?

4   A box of 60 coloured crayons contains a mixture of colours, 10 of which are red. If one crayon is removed at random, what is the probability that it is not red?

5   A number is chosen at random from the first 10 whole numbers. What is the probability that it is not exactly divisible by 3?

6   One letter is chosen at random from the letters of the word ALPHABET. What is the probability that it is not a vowel?

7   In a raffle, 500 tickets are sold. If you buy 20 tickets, what is the probability that you will not win first prize?

8   If you throw an ordinary fair six-sided die, what is the probability that you will not get a score of 5 or more?

9   There are 200 packets hidden in a lucky dip. Five packets contain $100 and the rest contain 1 c. What is the probability that you will not draw out a packet containing $100?

10 When an ordinary pack of playing cards is cut, what is the probability that the card showing is not a picture card? (For this question the picture cards are the jacks, queens and kings.)

11 A letter is chosen at random from the letters of the word SUCCESSION. What is the probability that the letter is
    a   N   b   S
    c   a vowel   d   not S?

12 A card is drawn at random from an ordinary pack of playing cards. What is the probability that it is
    a   an ace   b   a spade
    c   not a club   d   not a seven or an eight?

13 A bag contains a set of snooker balls (i.e. 15 red and 1 each of the following colours: white, yellow, green, brown, blue, pink and black). What is the probability that one ball removed at random is
    a   red   b   not red
    c   black   d   not red or white?

14 There are 60 cars in the car park. Of the cars, 22 are American made, 24 are Japanese made and the rest are European. What is the probability that the first car to leave is

  a Japanese   b not American

  c European   d Korean?

Expectation

Sometimes it is useful to be able to estimate how often an event *might* happen. This is called expectation.

For example, Sue is organising a game at a fête. The game involves rolling a die. When a six shows, you win a prize.
Sue needs to estimate the number of prizes that will be won.

On one turn, the probability of winning is $\frac{1}{6}$. So there will be about 1 win in every 6 turns.

Sue estimates that there will be about 300 turns. So there will be about $\frac{1}{6}$ of 300 wins.

Now $\frac{1}{6}$ of 300 = 300 ÷ 6 = 50 so Sue will need about 50 prizes.

$$\begin{pmatrix} \text{The number of times} \\ \text{that an event is likely} \\ \text{to happen} \end{pmatrix} = \begin{pmatrix} \text{probability that it} \\ \text{will happen once} \end{pmatrix} \times \begin{pmatrix} \text{the number of} \\ \text{times it is tried} \end{pmatrix}$$

When we flip a fair coin, the probability that it will show a head is $\frac{1}{2}$.

If we flip this coin 20 times, we expect to get about $\frac{1}{2} \times 20$ heads, i.e. 10 heads.

What we expect to get is an *estimate*. It is not the same as what we will get.

Suppose we flip a coin 20 times and get 15 heads. This is more than the 10 heads we expect. This can happen by chance, or it could be that the coin is biased. Biased means that some outcomes are more likely than others.

Exercise 21f

1 A fair coin is flipped 100 times.
   Write down the number of heads expected.

2 A fair ordinary die is rolled 60 times.
   Find the number of times you expect it to show 1.

Fair means the same as unbiased, i.e. any outcome is as likely to happen as any other possible outcome.

3   This spinner is fair. It is spun 100 times.
     Work out the number of times you expect it to show 4.

4   This spinner is fair. It is spun 40 times.
     a   Find the number of times you expect it will show 1.
     b   Work out the number of times you expect that it will show 3.

5   This fair spinner is spun 50 times.
     How often do you expect it will show an even number?

6   An ordinary fair unbiased die is rolled 90 times.
     a   Work out the number of times you expect it to show 6.
     b   Work out the number of times you expect it to show an odd number.
     c   Work out the number of times you expect it to show a number less than 3.

7   Jamestown airport has 500 flights leaving each day.

     The probability that a plane is delayed is $\frac{1}{25}$.

     How many delayed flights are expected on one day?

8   Derek rolls two ordinary fair six-sided dice 360 times.

     The probability that they show a double six is $\frac{1}{36}$.

     Work out the number of double sixes Derek is likely to get.

9   The probability of winning a prize on a game of chance is $\frac{1}{50}$.
     500 people have a go at this game. Find an estimate for the number of
     prizes that were won.

10   The probability that a lift will break down each time it is used is $\frac{1}{1000}$.
     The lift is used about 5000 times each year. Estimate the number of
     times that it is likely to break down.

11   Sam goes to work by bus. The probability that his bus is late is $\frac{1}{8}$.

     Sam works 200 days each year. Estimate the number of these days
     when Sam's bus is late.

12   One card is picked at random from an ordinary pack of playing cards.
     a   Write down the probability that it is an ace.
     b   Joe has 40 packs of ordinary playing cards.
          One card is picked at random from each pack.
          Estimate the number of aces that are picked.

13   a   Write down the probability of getting a head on one flip of a fair coin.
     b   Work out the number of heads you expect to get if you flipped the
          coin 60 times.
     c   Angela said 'If you flip a fair coin 10 times, you will get 5 heads.'
          Is Angela correct? Give a reason for your answer.
     d   Sam tossed one coin 100 times. He got 10 heads.

Sam said 'This coin is biased.'

Is Sam correct? Give a reason for your answer.

14 Harry plays a game of chance. The probability of winning is $\frac{1}{10}$. Each
go at the game costs $20 and the prize is $100.

Harry plays this game 50 times.

a How much does Harry spend on the game?

b How many times is Harry likely to win?

c How much money is Harry likely to win?

d Is Harry likely to lose money or make
money on this game?

Give a reason for your answer.

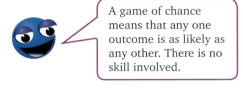

A game of chance
means that any one
outcome is as likely as
any other. There is no
skill involved.

15 The probability of winning a prize of $2500 from a scratch card is $\frac{1}{50}$.

Scratch cards cost $100 each.

Tom buys 100 cards.

a Work out how many times Tom is likely to win.

b Work out how much Tom is likely to win.

c Calculate the difference between the money Tom spends on
scratch cards and his likely winnings.

16 The probability that the same symbol shows in each of the three
windows on this slot machine is $\frac{1}{64}$.

When this happens, the machine pays out $1000.

Otherwise it pays out nothing.

a Kate plays 128 turns.

Write down the number of times that Kate is likely to win.

b Work out the amount of money that Kate is likely to win.

c Each turn costs $20.

Work out how much 128 turns costs Kate.

d Is Kate likely to make a profit?

Give a reason for your answer.

17 Rachel plays a game with four fair ordinary dice.

The four dice are rolled.

When at least one six shows, Rachel wins $100.

The probability that this will happen is 0.52.

When no sixes show Rachel loses $100.

She plays this game 100 times.

Is she likely to make a profit from this game?

Give a reason for your answer.

Investigation

This question should be done as a class exercise.

A company runs the following promotional offer with tubes of sweets.

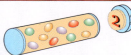

Each tube lid has a number 1, 2, 3, 4,
or 6 printed inside.

Collect 4 lids with the same number
and get a free T-shirt

What are the chances of getting four numbers the same if you just
buy four tubes?

What is the most likely number of tubes that you need to buy to
collect enough lids?

1 We could answer the questions by collecting evidence, i.e. buying tubes
of these sweets until we have four lids with the same number on them,
and repeating this until we think we have reliable results. What are the
disadvantages of this?

2 We can avoid the disadvantages of having to buy tubes
of these sweets by simulating the situation as follows:

First assume that any one of the numbers
1 to 6 is equally likely to be inside a lid.

Now throw a die to simulate the number we would
get if we bought a tube, and carry on throwing it until
you get four numbers the same. We will need several
tally charts to keep a record of the number of throws
needed on each occasion.

Score	Tally
1	|
2	|||
3	|
4	||
5	||||
6	|
Total	12

This shows the start of the simulation.

Number of throws needed to get 4 numbers the same	4	5	6	7	8	9	10	11	12	13	14	15	16	...
Tally									|					

In this chapter you have seen that...

✔ the probability that an event A will happen

$$P(A) = \frac{\text{the number of ways that the event can happen}}{\text{the total number of equally likely outcomes}}$$

✔ the probability that an event will happen lies between 0 and 1. It is 0 when the event is impossible and it is 1 when the event is certain

✔ the probability that event A will not happen is $1 - P(A$ does happen$)$

✔ an estimate for the number of times an event will happen = (probability that it will happen once) × (the number of times it is tried).

REVIEW TEST 3: CHAPTERS 15–21

In questions **1** to **13** choose the letter for the correct answer.

1 The floor area of a rectangular room measuring 8 m by 4 m is

    A 24 m²   **B** 32 m²   **C** 48 m²   **D** 64 m²

2 Which of these nets will make a closed cube?

    A   **B**   **C**   **D**

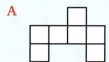 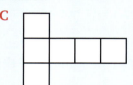

3 The object A is reflected in the x-axis.
The image of A is

    A a   **B** b   **C** c   **D** d

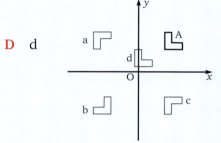

4 Given $3x - 8 = 4$. $x =$

    A 0   **B** $1\frac{1}{3}$ **C** 3   **D** 4

5 Given that $P = 2a + b$, the value of P when $a = 2\frac{1}{2}$ and $b = 4$ is

    A 1   **B** $6\frac{1}{2}$ **C** 9   **D** 13

6 The area of a rectangle is 4.5 cm². The length of the rectangle is 9 cm.
The width is

    A 0.05 cm   **B** 0.5 cm   **C** 5 cm   **D** 50 cm

Questions **7** and **8** refer to this diagram:

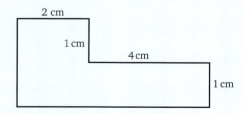

7   The perimeter of this shape is
    A   8 cm   **B**   12 cm   **C**   16 cm   **D**   20 cm

8   The area of this shape is
    A   6 cm²   **B**   7 cm²   **C**   8 cm²   **D**   16 cm²

9   A pie chart shows the age distribution of persons in Caricom
    countries. Twenty-five per cent of the people are in the age range
    'over 60 years'. What is the size of the angle representing this group?
    A   25°   **B**   60°   **C**   75°   **D**   90°

10 $3x - 2y + 7x + 9y$ simplifies to
    A $10x + 7y$ **B** $x + 16y$ **C** $10x + 11y$ **D** $17x + y$

11 I start with n pencils and then buy m pencils. The formula for N the
    number of pencils I now have is
    A $N = n \times m$ **B** $N = n - m$ **C** $m = N + n$ **D** $N = n + m$

12 When you roll an ordinary fair die, the probability that you will not
    get a score less than 5 is
    A $\frac{1}{6}$ **B** $\frac{1}{3}$ **C** $\frac{2}{3}$ **D** $\frac{5}{6}$

13 In a class of 36 students, 9 walk to school, 10 cycle, 11 come by
    bus, 5 are driven by a relative and 1 is driven by a chauffeur. A pie
    chart shows this information. The slice that represents the number of
    children who walk to school has an angle at its centre of
    A   9°   **B**   45°   **C**   90°   **D**   180°

14 **a**   If n is an even integer, what is the next even number above n?
    b   If you are m years old, what must be your father's present age if he
          is 6 years more than three times your age?
    c   Solve the equation $3x - 9 = 2x - 2$

15 **a** Complete the following table showing the masses of students in a class.

Mass (kg)	Tally	Number
41–45	|	
46–50	|||	
51–55	Ⅲ⅃ Ⅲ⅃ |	11
56–60	Ⅲ⅃ |||	
61–65	||||	4
66–70	|||	

b Draw a bar chart showing the information.

16 Find the area of this figure.

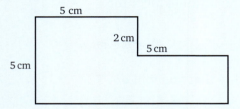

17 The marks out of 10 in a maths test were

5, 7, 4, 5, 7, 2, 5, 3, 9, 10, 10, 5, 8, 4, 10, 1, 9, 7, 5, 8, 2, 4.

Make a frequency table showing these marks.

REVIEW TEST 4: CHAPTERS 1–21

Choose the letter for the correct answer.

1 The smallest prime number is

 A 1   **B** 2   **C** 7   **D** 9

2 $\frac{0.3}{100}$ has the same value as

 A 0.003%   **B** 0.03%   **C** 0.3%   **D** 3%

3 X = {odd numbers from 2 to 5}

 Y = {prime numbers from 2 to 5}

 The element not common to both X and Y is

 A 2   **B** 3   **C** 4   **D** 5

4 From the numbers 1 to 10, the largest prime number exceeds the smallest prime number by

 A 8   **B** 7   **C** 6   **D** 5

5 An item bought for $120 is sold to make a profit of 20%. The selling price is

 A $96   **B** $125   **C** $140   **D** $144

6 In an exam marked out of 120 marks, a student requires a minimum of 80% to attain a grade 'A'. The minimum mark required for a grade 'A' is

 A 80   **B** 96   **C** 100   **D** 120

7 Two sets of 'flashing lights' start together and then flash at intervals of 5 seconds and 6 seconds respectively. They will flash together at intervals of

 A 6 seconds   **B** 11 seconds   **C** 15 seconds   **D** 30 seconds

8 The square and the rectangle shown in the diagram above have the same area. The width of the rectangle is

Rectangle	Square
16 cm	8 cm

 A 8 cm   **B** 6 cm   **C** 4 cm   **D** 2 cm

9 The length of the line segment joining (3, 4) and (8, 4) is

 A 3   **B** 4   **C** 5   **D** 11

10 A box contains 9 bulbs of which 3 are defective. If one bulb is chosen at random then the probability that it is not defective is

A $\frac{1}{9}$ B $\frac{1}{3}$ C $\frac{2}{3}$ D $\frac{8}{9}$

11 $R = \{$Even numbers from 1 to 10$\}$
$I = \{$whole numbers from 1 to 10$\}$
The diagram illustrating R and I is

A 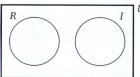   B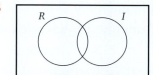

C 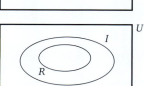   D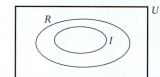

12 $\frac{2}{7} \div \frac{6}{7} =$

A $\frac{1}{3}$ B $\frac{3}{7}$ C $\frac{7}{3}$ D 3

13 A liquid weighs 800 g per litre. The number of litres that weigh 1 kg is

A 1.25   B 1.8   C 12.5   D 8

14 The pie chart shows how $10 000 000 collected by a charity for earthquake victims is spent.
The amount spent on clothes is

A $2 500 000   B $5 000 000
C $7 500 000   D $10 000 000

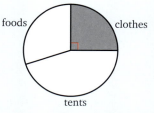

15 $0.1 \div 0.01 =$

A 10 000   B 1000   C 100   D 10

16 $\frac{5}{8}$ written as a decimal is

A 0.5   B 0.625   C 0.8   D 0.875

17 The coordinates of the point N are

A (0, 2) B (2, 0)

C (4, 0) D (0, 4)

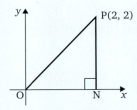

18 In the quadrilateral PQRS, the diagonals PR and QS intersect at right angles at O with OQ = OS. The quadrilateral can best be described as a

A Rhombus B Trapezium

C Parallelogram D Kite

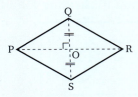

19 Given that $S = 5t - 8u$, the value of S when $t = 20$ and $u = 10$ is

A 2 B 8 C 10 D 20

20 $3(x - 4) = 2x - 3$, $x =$

A 1 B 9 C 10 D 15

21 $\frac{2}{3} + \frac{1}{6} =$

A $\frac{2}{18}$ B $\frac{3}{9}$ C $\frac{3}{6}$ D $\frac{5}{6}$

22 The area of this figure is

A 9 cm² B 8 cm²

C 5 cm² D 4 cm²

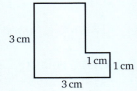

23 This solid is best described as

A a cube B a cylinder

C a prism D a cuboid

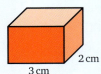

24 $2(x + 3) = 9$, x is

A $1\frac{1}{2}$ B 3 C 6 D $7\frac{1}{2}$

25 $6x - 4 + 10 - 6x =$

A x B 6 C 10 D $12 - 2x$

26 Josh and Emma share $60. Josh gets $\frac{2}{5}$ of the money. Josh gets the amount

    A $20   **B** $24   **C** $30   **D** $60

27 P = {2, 4, 6, 8}, Q = {1, 2, 3, 4}, P∩Q =

    A {4}   **B** {2, 4}   **C** { }   **D** {1, 2, 3, 4, 6, 8}

28 Mark will toss a coin 200 times. The number of heads expected is

    A 200   **B** 100   **C** 50   **D** 2

29 This shape is called a

    A triangle   **B** square

    C parallelogram   **D** rectangle

30 160 miles in kilometres is about

    A 320 km   **B** 256 km   **C** 240 km   **D** 100 km

Answers

CHAPTER 1

Exercise 1a   page 2
1 **a** 63 **b** 49 **c** 707 **d** 327
2 **a** 819 **b** 8008 **c** 6067 **d** 15 234
3 **a** fifty-six
  b seventy-nine
  c four hundred and nine
  d one hundred and eighty-seven
  e seven hundred and thirty-four
4 **a** three hundred and thirty
  b four hundred and twenty-six
  c nine thousand four hundred and eighty-eight
  d six thousand five hundred and ninety-three
  e seven thousand and sixty-five
5 **a** 5 **b** 6
6 **a** 9 **b** 0 **c** 8
7 **a** 43, 55, 57, 63
  b 27, 31, 49, 83
  c 77, 104, 293, 308
8 **a** 506, 560, 605, 650
  b 98, 845, 1088, 8876
  c 2033, 2303, 3032, 3302
9 **a** 40 **b** 400 **c** 4
  d 4000 **e** 40 000
10 **a** 600 **b** 6 **c** 6000 **d** 60 000
11 **a** 974 **b** 479
12 **a** 540 **b** 405
13 2379
14 6532
15 5017
16 5864
17 **a** 2440 **b** 2040
18 543, 534, 453, 435, 354, 345
19 267, 276, 627, 672, 726, 762

Exercise 1b   page 4

1 10	8 19	15 33	22 17	29 26
2 11	9 20	16 18	23 20	30 32
3 14	10 27	17 25	24 33	31 26
4 15	11 15	18 32	25 30	32 26
5 17	12 17	19 39	26 21	33 40
6 24	13 27	20 32	27 21	34 37
7 24	14 27	21 24	28 19	35 39

Exercise 1c   page 6

1 79	20 1966
2 97	21 183
3 65	22 177
4 308	23 202
5 259	24 1252
6 399	25 2783
7 882	26 2062
8 2039	27 1267
9 991	28 2764
10 2292	29 5936
11 549	30 7525
12 1835	31 1693
13 9072	32 1382
14 21 829	33 1896
15 16 244	34 5230
16 112	35 4095
17 158	36 581
18 242	37 509
19 797	38 857

39 1087	43 2226
40 1832	44 3569
41 2892	45 11 932
42 6779	

Exercise 1d   page 7
1 $159 **3** 88
2 $160 **4** $52 990
5 **a** 261 **b** 302
  c 3056 **d** 1300
6 **a** three hundred and twenty-four
  b five thousand two hundred and eight
  c one hundred and fifty
  d one thousand five hundred
7 787 **10** 50 min
8 77 cm **11** 4957
9 $5300 **12** $1023

Exercise 1e   page 9

1 11	5 7	9 8	13 11	17 5
2 12	6 12	10 6	14 8	18 6
3 14	7 15	11 13	15 10	19 14
4 5	8 9	12 3	16 4	20 8

Exercise 1f   page 9

1 211	10 186	19 279	28 189
2 551	11 470	20 149	29 703
3 406	12 354	21 8	30 676
4 218	13 287	22 2828	31 4077
5 73	14 178	23 4823	32 1048
6 141	15 187	24 6615	33 77
7 406	16 136	25 575	34 192
8 126	17 713	26 3344	35 4195
9 126	18 255	27 1524	36 1644

Exercise 1g   page 10

1 $4030	4 89	7 213	10 19 cm
2 464	5 287	8 48	
3 85	6 6483	9 6623 m	

Exercise 1h   page 11

1 6	3 7	5 9	7 2	9 9
2 5	4 4	6 4	8 7	10 7

Exercise 1i   page 12

1 17	11 0	21 73
2 5	12 25	22 20
3 2	13 0	23 104
4 20	14 67	24 7
5 30	15 83	25 29
6 28	16 50	26 597
7 13	17 0	27 19
8 3	18 39	28 129
9 6	19 0	29 250
10 4	20 95	30 65

Exercise 1j   page 13

1 $43	5 144	9 $9
2 72	6 69 kg	10 Jan 1; Dec 31
3 80 cm	7 17	
4 318	8 45	

Exercise 1k   page 15

1 8	3 5
2 15	4 63

5 1
6 4
7 23
8 16
9 7
10 0
11 8
12 3
13 8
14 12
15 14
16 5
17 16
18 38
19 10
20 20
21 250, 257
22 60, 56
23 210, 209
24 510, 507
25 330, 334
26 40, 38
27 370, 366

28 260, 264
29 180, 176
30 770, 777
31 60, 58
32 20, 16
33 160, 163
34 160, 154
35 150, 148
36 40, 42
37 280, 284
38 230, 229
39 370, 362
40 160, 160
41 370, 360
42 210, 206
43 230, 227
44 250, 251
45 330, 328
46 290, 293
47 250, 250
48 300, 291
49 180, 170
50 360, 353

13 60 000
14 300 000
15 240 000
16 300 000, 244 326
17 12 000, 11 136
18 12 000, 10 192
19 36 000, 34 225
20 16 000, 18 768
21 7200, 7098
22 6000, 8750
23 30 000, 32 406
24 30 000, 30 012
25 7200, 6612
26 40 000, 42 692
27 45 000, 42 987
28 50 000, 46 657
29 600 000, 579 424

30 300 000, 298 717
31 5600, 5382
32 45 000, 40 091
33 54 000, 51 888
34 1000, 846
35 6000, 6076
36 45 000, 40 281
37 24 000, 22 222
38 560 000, 563 997
39 25 000, 23 124
40 35 000, 35 972
41 24 000, 23 458
42 200 000, 231 548
43 480 000, 465 234
44 4 900 000, 5 053 014
45 350 000, 346 320

CHAPTER 2

Exercise 2a   page 19
1 46
2 126
3 104
4 304
5 290
6 93
7 100
8 144
9 144
10 415
11 141
12 324
13 126
14 588
15 324
16 292
17 162
18 132
19 536
20 657
21 294
22 168
23 224
24 243
25 608
26 2456
27 768
28 388
29 1989
30 844
31 2859
32 1632
33 2628
34 2184
35 852
36 2565
37 3174
38 5142
39 3486
40 5211
41 4606
42 2989
43 6784
44 5931
45 5236
46 5552
47 1652
48 5157

Exercise 2b   page 21
1 270
2 8200
3 360
4 1080
5 256 000
6 540
7 24 600
8 2040
9 7800
10 2800
11 29 200
12 3480
13 6630
14 88 900
15 146 000
16 35 100
17 9420
18 23 600
19 6160
20 70 000
21 48 720
22 54 000
23 38 920
24 243 000
25 35 100
26 42 800
27 19 200
28 8800
29 19 000
30 59 920

Exercise 2c   page 22
1 672
2 559
3 1290
4 567
5 1428
6 1558
7 2782
8 4346
9 7844
10 3204
11 7712
12 40 086
13 398 793
14 35 028
15 112 893
16 107 520
17 39 934
18 70 952
19 37 814
20 565 915
21 86 172
22 56 648
23 169 422
24 191 430
25 1 438 200
26 36 575
27 337 500
28 453 750
29 915 264
30 1 203 000

Exercise 2d   page 23
1 2400
2 900
3 3200
4 1500
5 9000
6 1200
7 1200
8 3600
9 3000
10 15 000
11 18 000
12 24 000

Exercise 2e   page 24
1 8188
2 10 896
3 272
4 840
5 22 500
6 1428
7 2592
8 420
9 792
10 672

Exercise 2f   page 25
1 29
2 14
3 6
4 19
5 18
6 48 r1
7 14 r3
8 20 r3
9 23
10 13 r4
11 9 r6
12 12 r1
13 13
14 2 r3
15 13
16 27
17 213
18 274
19 201 r2
20 124 r1
21 171
22 231
23 103
24 71 r3
25 24
26 32 r6
27 81 r3
28 85
29 121 r5
30 140 r2
31 1167
32 440 r3
33 2414 r1
34 351 r3
35 428
36 1067 r3
37 1479 r4
38 2193
39 1214
40 198 r6
41 287
42 183
43 354 r3
44 1727 r2
45 1501

Exercise 2g   page 26
1 25 r6
2 8 r7
3 1 r96
4 27 r83
5 4 r910
6 5 r7
7 18 r6
8 278 r1
9 9 r426
10 85 r12
11 30 r77
12 5 r704

Exercise 2h   page 27
1 12 r14
2 52 r9
3 18 r1
4 34 r12
5 20 r14
6 8 r11
7 35 r0
8 16 r13
9 16 r21
10 21 r4
11 28 r13
12 22 r20
13 215 r9
14 348 r7
15 246 r28
16 456 r1
17 127 r22
18 86 r28
19 75 r0
20 120 r21
21 221 r0
22 135 r24
23 236 r0
24 217 r15
25 304 r19
26 573 r7
27 96 r28
28 64 r8
29 202 r22
30 89 r24
31 200 r13
32 65 r14
33 83 r29
34 146 r34
35 77 r9
36 469 r1
37 2 r33
38 107 r17
39 111 r13
40 190 r20
41 25 r0
42 111 r5
43 90 r30
44 200 r0
45 11 r6
46 20 r10
47 20 r4
48 42 r38
49 7 r87
50 26 r15
51 24 r65
52 32 r200
53 12 r6
54 56 r91
55 25 r75
56 20 r110
57 6 r142
58 74 r44
59 27 r109
60 22 r152

Exercise 2i   page 28

1 18	**9** 7	**17** 3	**25** 6	**33** 12
2 0	**10** 21	**18** 13	**26** 8	**34** 13
3 12	**11** 9	**19** 26	**27** 10	**35** 32
4 19	**12** 17	**20** 6	**28** 8	**36** 9
5 0	**13** 2	**21** 8	**29** 5	**37** 16
6 5	**14** 5	**22** 22	**30** 9	**38** 14
7 22	**15** 1	**23** 13	**31** 21	**39** 14
8 7	**16** 10	**24** 17	**32** 14	**40** 30

Exercise 2j   page 30

1 2	**7** 49	**13** 17	**19** 4	**25** 10
2 56	**8** 2	**14** 2	**20** 36	**26** 1
3 9	**9** 45	**15** 11	**21** 45	**27** 4
4 14	**10** 2	**16** 7	**22** 6	**28** 25
5 15	**11** 17	**17** 30	**23** 14	**29** 1
6 8	**12** 3	**18** 1	**24** 0	**30** 18

Exercise 2k   page 31

1 6 and $20 over	**11** 76
2 $180	**12** $46
3 14	**13** $200
4 $45	**14** 90
5 $12	**15** 840 cm
6 $15	**16** $9, $18, $33
7 150 miles	**17** $237
8 74	**18** $4000
9 $325	**19** 225 275
10 16 and 2 kg over	**20** 54 (one not full)

21 67
22 1831 or 1832 depending on her birth date

23 26	**27** 15
24 86 years	**28** $265
25 600 m	**29** 34
26 12 min	**30** 2 h, 25 min

Exercise 2l   page 33

1

8	1	6
3	5	7
4	9	2

2

4	9	2
3	5	7
8	1	6

3

2	14	7	11
15	3	10	6
9	5	16	4
8	12	1	13

5 **6** **7**

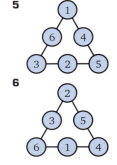

8 (i)

s	10	11	12
υ	9	12	15
$3s-\upsilon$	21	21	21

(ii) 21, 21, 21
(iii) equal

9 9, 11	**16** 81, 243
10 13, 16	**17** 36, 49
11 4, 2	**18** 10 000, 100 000
12 17, 21	**19** 45, 36
13 32, 64	**20** 19, 23
14 15, 18	**21** 37, 50
15 4, 2	

Investigations

1 $1+3+5+7+9$ $= 25 = 5 \times 5$
$1+3+5+7+9+11$ $= 36 = 6 \times 6$
$1+3+5+7+9+11+13$ $= 49 = 7 \times 7$
a 64   **b** 400

2 $2+4+6+8+10$ $= 30 = 5 \times 6$
$2+4+6+8+10+12$ $= 42 = 6 \times 7$
$2+4+6+8+10+12+14$ $= 56 = 7 \times 8$
12

3
1		5		10		10		5		1
1		6		15		20		15		6		1
1		7		21		35		35		21		7		1

4 35

5 $1-3+5-7+9-11+13-15+17 = 9$
$1-3+5-7+9-11+13-15+17-19+21 = 11$
$1-3+5-7+9-11+13-15+17-19+21-23+25 = 13$

Exercise 2m   page 37

1 4, 9, 36
2 8, 6, 14, 15
3 2×6, 3×4
4 2×9, 3×6
5 2×18, 3×12, 4×9 or 6×6
6

15, 21, 28
7 36, 45, 55
8 1, 4, 9, 16; square numbers
9 1, 4, 9, 16; square numbers
10 triangular numbers
11 a 1, 4, 9
   b 4, 6, 8, 9, 10, 12
   c 1, 3, 6, 10
12 a 25, 36
   b 24, 25, 26, 27, 28, 30, 32, 33, 34, 35, 36, 38, 39, 40
   c 28, 36

Exercise 2n   page 38

1 1005	**6** 1018
2 17	**7** 242
3 684	**8** 7
4 28	**9** 2 ($10 over)
5 6608	**10** $16

Exercise 2p   page 38

1 870	**6** 118
2 54	**7** 50
3 672	**8** 37
4 9 r7	**9** 7 (3 left)
5 29	**10** 5

Exercise 2q   page 39

1 2304	**5** 277 r8	**9** 35, 45
2 263	**6** 393 r3	**10** 33
3 413	**7** 260	
4 3392	**8** 19 r133	

Exercise 2r   page 39

1 3133	**4** 4544	**7** 3
2 169	**5** 278 r1	**8** 132
3 8200	**6** 713	**9** $37

10 1, 4, 9; square numbers

CHAPTER 3

Exercise 3a   page 41

1 a {foreign cars}
  b {pupils in my class}
  c {subjects/study at school}
  d {furniture in this room}

Exercise 3b   page 42

1 the last four letters of the alphabet
2 the months whose names begin with J
3 the 6th to 8th months of the year
4 the Windward islands
5 four islands in the Leeward islands
6 even numbers less than 13
7 the first six whole numbers
8 the first six prime numbers
9 the whole numbers from 45 to 50 inclusive
10 multiples of 5 from 15 to 35 inclusive
11 {boys' names}
12 {outerwear}
13 {breakfast cereals}
14 {Caribbean plants}
15 {plays by Shakespeare}
16 {11, 12, 13, 14, 15}
17 {a, b, c, d, e, f, g, h}
18 {a, c, e, h, i, m, s, t}
19 {Grenada, St Vincent, St Lucia, Dominica}
20 {Antigua, St Kitts, Montserrat, Guadeloupe}
22 {Arctic, Antarctic, Pacific, Atlantic, Indian}
24 {2, 3, 5, 7, 11, 13, 17, 19}
25 {2, 4, 6, 8, 10, 12, 14, 16, 18}
26 {21, 23, 25, 27, 29}
27 {12, 15, 18, 21, 24, 27, 30}
28 {21, 28, 35, 42, 49}
29 {St George's, Kingstown, Castries, Roseau}
30 {Guyana, Trinidad, Jamaica, Barbados,
   Grenada, St Vincent, St Lucia, Dominica,
   Antigua, St Kitts, Montserrat}

Exercise 3c   page 43

1 apple ∈ {fruit}
2 shirt ∈ {clothing}
3 dog ∈ {domestic animals}
4 geography ∈ {school subjects}
5 carpet ∈ {floor coverings}
6 hairdressing ∈ {occupations}

Exercise 3d   page 44

1 orange ∉ {animals}
2 cat ∉ {fruit}
3 table ∉ {trees}
4 shirt ∉ {subjects/study}
5 Anne ∉ {boys' names}
6 chisel ∉ {buildings}
7 cup ∉ {bedroom furniture}
8 Mercedes ∉ {Japanese cars}
9 aeroplane ∉ {foreign countries}
10 curry ∉ {breeds of dogs}
11 porridge ∈ {breakfast cereals}
12 electricity ∉ {building materials}
13 water ∉ {metals}
14 spider ∈ {living things}
15 Saturday ∈ {days of the week}
16 snapper ∈ {fish}

17 August ∉ {days of the week}
18 Spain ∈ {European countries}
19 Brazil ∉ {Asian countries}
20 Football is a team game.
21 Shoes are not a beverage.
22 Hockey is not an electrical appliance.
23 A needle is a metal object.
24 Susan is not a boys' name.

Exercise 3e   page 46

1 Yes	**2** No	**3** Yes	**4** Yes	**5** Yes

Exercise 3f   page 46

2 a, c, d

Exercise 3g   page 47

1 {x, y}, {w, x}, {y, z}, {w, y}, {x, z}, {w, z}
2 {A, D}, {A, D, C, B}, {A, D, C}, {A, D, B}
3 $A = \{1, 3, 5, 7, 9\}$ $B = \{2, 4, 6, 8, 10\}$ $C = \{2, 3, 5, 7\}$
5 a, b, c, d

Exercise 3h   page 48

1 {whole numbers}
2 {letters of the alphabet}
3 {rivers}   **5** {cats}
4 {pupils}   **6** {birds}
13 {geometrical instruments}
14 {dwellings}
15 {cars}
16 {footwear}
17 {sportsmen and sportswomen}

Exercise 3i   page 51

1 {Peter, James, John, Andrew, Paul}
2 {3, 4, 6, 8, 9, 12, 16}
3 {a, b, c, d, e, i, o, u}
4 {a, b, c, x, y, z}
5 {p. q. r. s. t} = A
6 {1, 2, 3, 5, 7}
7 {5, 6, 7, 8, 10, 11, 12, 13}
8 {1, 2, 3, 4, 5, 6, 10, 12}
9 {a, c, h, 1, m, o, r, s}
10 {a, b, c, e, g, h, i, l, m, r, t}

Exercise 3j   page 52

1
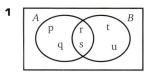

2
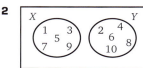

3
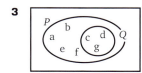

4
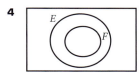

5

6

7

8

9

10

14 1×96, 2×48, 3×32, 4×24, 6×16, 8×12
15 1×100, 2×50, 4×25, 5×20, 10×10
16 1×108, 2×54, 3×36, 4×27, 6×18, 9×12
17 1×120, 2×60, 3×40, 4×30, 5×24, 6×20, 8×15, 10×12
18 1×135, 3×45, 5×27, 9×15,
19 1×144, 2×72, 3×48, 4×36, 6×24, 8×18, 9×16, 12×12
20 1×160, 2×80, 4×40, 5×32, 8×20, 10×16

Exercise 4b   page 57
1 {1, 2, 3, 6, 9, 18}
2 {1, 2, 4, 5, 10, 20}
3 {1, 2, 3, 4, 6, 8, 12, 24}
4 {1, 3, 9, 27}
5 {1, 2, 3, 5, 6, 10, 15, 30}
6 {1, 2, 3, 4, 6, 9, 12, 18, 36}
7 {1, 2, 4, 5, 8, 10, 20, 40}
8 {1, 3, 5, 9, 15, 45}
9 {1, 2, 3, 4, 6, 8, 12, 16, 24, 48}
10 {1, 2, 3, 4, 5, 6, 10, 12, 15, 20, 30, 60}
11 {1, 2, 4, 8, 16, 32, 64}
12 {1, 2, 3, 4, 6, 8, 9, 12, 18, 24, 36, 72}
13 {1, 2, 4, 5, 8, 10, 16, 20, 40, 80}
14 {1, 2, 3, 4, 6, 8, 12, 16, 24, 32, 48, 96}
15 {1, 2, 4, 5, 10, 20, 25, 50, 100}
16 {1, 2, 3, 4, 6, 9, 12, 18, 27, 36, 54, 108}
17 {1, 2, 3, 4, 5, 6, 8, 10, 12, 15, 20, 24, 30, 40, 60, 120}
18 {1, 3, 5, 9, 15, 27, 45, 135}
19 {1, 2, 3, 4, 6, 8, 9, 12, 16, 18, 24, 36, 48, 72, 144}
20 {1, 2, 4, 5, 8, 10, 16, 20, 32, 40, 80, 160}

Exercise 4c   page 57
1 21, 24, 27, 30, 33, 36, 39
2 20, 25, 30, 35, 40, 45
3 28, 35, 42, 49, 56
4 55, 66, 77, 88, 99
5 26, 39, 52, 65

Exercise 4d   page 58
1 2, 3, 5, 7, 11, 13
2 23, 29
3 31, 37, 41, 43, 47
4 5, 19, 29, 61
5 41, 101, 127
6 a F   **b** F   **c** T   **d** T   **e** F

Exercise 4e   page 59
1 yes   **5** no   **9** yes
2 no   **6** yes   **10** yes
3 yes   **7** yes   **11** yes
4 yes   **8** no

Exercise 4f   page 60
1 3   **8** 6   **15** 17
2 8   **9** 6   **16** 5
3 12   **10** 8   **17** 4
4 14   **11** 25   **18** 15
5 7   **12** 22   **19** 2
6 7   **13** 42   **20** 2
7 3   **14** 13

Exercise 4g   page 61
1 15   **5** 36   **9** 36
2 24   **6** 60   **10** 108
3 15   **7** 48   **11** 36
4 36   **8** 60   **12** 168

Exercise 4h   page 62
1 50c   **4** 50 cm   **7** 480, 20
2 $216   **5** 90 sec   **8** 30 steps; 2
3 24 m   **6** 18

Exercise 3k   page 53
1 {6, 9}   **6** {3, 5, 7, 11}
2 {4, 12, 20}   **7** {8, 16}
3 {Alice, Bob}   **8** {1, 2}
4 {o}   **9** {e, t, w}
5 {cabbage, tomato}   **10** {m, e, i, r}

Exercise 3l   page 54
1 {3, 5, 7}   **6** {8, 16}
2 {Dino, John}   **7** {1, 2, 4}
3 {o, u}   **8** {t, i, n}
4 {oak, elm}   **9** {r, t, m, e}
5 {boxer}   **10** {3, 5, 7}

CHAPTER 4

Exercise 4a   page 57
1 1×18, 2×9, 3×6
2 1×20, 2×10, 4×5
3 1×24, 2×12, 3×8, 4×6
4 1×27, 3×9
5 1×30, 2×15, 3×10, 5×6
6 1×36, 2×18, 3×12, 4×9, 6×6
7 1×40, 2×20, 4×10, 5×8
8 1×45, 3×15, 5×9
9 1×48, 2×24, 3×16, 4×12, 6×8
10 1×60, 2×30, 3×20, 4×15, 5×12, 6×10
11 1×64, 2×32, 4×16, 8×8
12 1×72, 2×36, 3×24, 4×18, 6×12, 8×9
13 1×80, 2×40, 4×20, 5×16, 8×10

CHAPTER 5

Exercise 5a page 67

1 $\frac{1}{6}$	5 $\frac{2}{6}$	9 $\frac{1}{2}$	13 $\frac{3}{7}$
2 $\frac{3}{8}$	6 $\frac{7}{10}$	10 $\frac{3}{10}$	14 $\frac{2}{6}$
3 $\frac{1}{3}$	7 $\frac{1}{4}$	11 $\frac{5}{12}$	15 $\frac{4}{8}$
4 $\frac{5}{6}$	8 $\frac{3}{4}$	12 $\frac{1}{4}$	16 $\frac{1}{6}$

Exercise 5b page 68

1 a $\frac{1}{60}$ b $\frac{9}{60}$ c $\frac{30}{60}$ d $\frac{45}{60}$

2 $\frac{5}{7}$

3 $\frac{11}{30}$

4 $\frac{51}{365}$

5 $\frac{35}{100}$

6 $\frac{90}{500}$

7 $\frac{35}{180}$

8 $\frac{3}{31}$

9 $\frac{17}{61}$

10 $\frac{5}{21}$

11 $\frac{150}{500}$

12 $\frac{45}{120}$

13 $\frac{37}{3600}$

14 $\frac{35}{80}$

15 a $\frac{10}{32}$ b $\frac{8}{32}$ c $\frac{25}{32}$

16 $\frac{15}{40}, \frac{25}{40}$

17 a $\frac{20}{62}$ b $\frac{10}{62}$ c $\frac{48}{62}$

18 a $\frac{12}{37}$ b $\frac{8}{37}$ c $\frac{29}{37}$

19 a $\frac{9}{14}$ b $\frac{3}{14}$

Exercise 5c page 70

7 6	16 6	25 300
8 4	17 16	26 110
9 21	18 18	27 40
10 36	19 18	28 1000
11 18	20 30	29 90
12 4	21 10	30 8000
13 15	22 10	31 55
14 12	23 100	32 500
15 100	24 8	33 10000

34 a $\frac{12}{24}$ b $\frac{8}{24}$ c $\frac{4}{24}$ d $\frac{18}{24}$ e $\frac{10}{24}$ f $\frac{9}{24}$

35 a $\frac{6}{45}$ b $\frac{20}{45}$ c $\frac{27}{45}$ d $\frac{15}{45}$ e $\frac{42}{45}$ f $\frac{9}{45}$

36 a $\frac{27}{36}$ b $\frac{20}{36}$ c $\frac{6}{36}$ d $\frac{10}{36}$ e $\frac{21}{36}$ f $\frac{24}{36}$

37 a $\frac{12}{72}$ b $\frac{12}{16}$ c $\frac{12}{14}$ d $\frac{12}{15}$ e $\frac{12}{18}$ f $\frac{12}{24}$

38 b $\frac{2}{3} = \frac{6}{9}$ e $\frac{7}{10} = \frac{70}{100}$

Exercise 5d page 72

1 $\frac{1}{2}$	6 $\frac{3}{4}$	11 $\frac{3}{5}$	16 $\frac{4}{11}$
2 $\frac{5}{6}$	7 $\frac{3}{7}$	12 $\frac{3}{4}$	17 $\frac{2}{7}$
3 $\frac{4}{5}$	8 $\frac{5}{6}$	13 $\frac{3}{11}$	18 $\frac{5}{8}$
4 $\frac{2}{9}$	9 $\frac{3}{8}$	14 $\frac{5}{7}$	19 $\frac{3}{11}$
5 $\frac{3}{8}$	10 $\frac{6}{7}$	15 $\frac{5}{11}$	20 $\frac{7}{9}$

21 $\frac{9}{11}$

22 $\frac{2}{5}$

23 $\frac{3}{5}$

24 $\frac{5}{8}$

25 <

26 >

27 <

28 <

29 >

30 <

31 <

32 >

33 <

34 >

35 <

36 >

37 $\frac{7}{30}, \frac{1}{2}, \frac{3}{5}, \frac{2}{3}$

38 $\frac{4}{10}, \frac{5}{8}, \frac{13}{20}, \frac{3}{4}$

39 $\frac{1}{3}, \frac{1}{2}, \frac{7}{12}, \frac{5}{6}$

40 $\frac{3}{8}, \frac{2}{5}, \frac{1}{2}, \frac{7}{10}, \frac{17}{20}$

41 $\frac{1}{2}, \frac{17}{28}, \frac{5}{7}, \frac{3}{4}, \frac{11}{14}$

42 $\frac{2}{5}, \frac{1}{2}, \frac{14}{25}, \frac{3}{5}, \frac{7}{10}$

43 $\frac{5}{6}, \frac{7}{9}, \frac{2}{3}, \frac{11}{18}, \frac{1}{2}$

44 $\frac{3}{4}, \frac{7}{10}, \frac{13}{20}, \frac{3}{5}, \frac{1}{2}$

45 $\frac{3}{4}, \frac{17}{24}, \frac{2}{3}, \frac{7}{12}, \frac{1}{6}$

46 $\frac{4}{5}, \frac{23}{30}, \frac{11}{15}, \frac{7}{10}, \frac{2}{3}$

47 $\frac{3}{4}, \frac{5}{8}, \frac{19}{32}, \frac{1}{2}, \frac{7}{16}$

48 $\frac{5}{6}, \frac{4}{5}, \frac{3}{4}, \frac{7}{12}, \frac{1}{2}$

Exercise 5e page 75

1 $\frac{1}{3}$	7 $\frac{1}{3}$	13 $\frac{1}{5}$	19 $\frac{3}{5}$	25 $\frac{4}{5}$
2 $\frac{3}{5}$	8 $\frac{2}{3}$	14 $\frac{2}{5}$	20 $\frac{2}{5}$	26 $\frac{4}{7}$
3 $\frac{1}{3}$	9 $\frac{1}{2}$	15 $\frac{2}{7}$	21 $\frac{5}{9}$	27 $\frac{1}{3}$
4 $\frac{1}{2}$	10 $\frac{1}{4}$	16 $\frac{1}{3}$	22 $\frac{7}{11}$	28 $\frac{9}{11}$
5 $\frac{1}{3}$	11 $\frac{2}{7}$	17 $\frac{1}{2}$	23 $\frac{3}{4}$	29 $\frac{3}{4}$
6 $\frac{1}{2}$	12 $\frac{3}{10}$	18 $\frac{1}{5}$	24 $\frac{3}{11}$	30 $\frac{4}{5}$

Exercise 5f page 76

1 $\frac{3}{4}$	8 $\frac{2}{5}$	15 $\frac{1}{2}$	22 $\frac{15}{23}$	29 $\frac{9}{17}$
2 $\frac{1}{2}$	9 $\frac{11}{21}$	16 $\frac{9}{10}$	23 $\frac{8}{9}$	30 $\frac{12}{19}$
3 $\frac{5}{11}$	10 $\frac{1}{2}$	17 $\frac{3}{4}$	24 $\frac{2}{3}$	31 $\frac{13}{30}$
4 $\frac{10}{13}$	11 $\frac{11}{13}$	18 $\frac{11}{19}$	25 $\frac{4}{5}$	32 $\frac{5}{9}$
5 $\frac{19}{23}$	12 $\frac{4}{5}$	19 $\frac{1}{2}$	26 $\frac{2}{5}$	33 $\frac{1}{2}$
6 $\frac{3}{7}$	13 $\frac{6}{7}$	20 $\frac{2}{5}$	27 $\frac{23}{31}$	34 $\frac{25}{99}$
7 $\frac{3}{5}$	14 $\frac{9}{17}$	21 $\frac{6}{11}$	28 $\frac{11}{14}$	

Exercise 5g page 77

1 $\frac{13}{15}$	10 $\frac{41}{42}$	19 $\frac{8}{9}$	28 $\frac{11}{12}$
2 $\frac{23}{40}$	11 $\frac{82}{99}$	20 $\frac{13}{18}$	29 $\frac{6}{7}$
3 $\frac{11}{30}$	12 $\frac{47}{90}$	21 $\frac{13}{20}$	30 1
4 $\frac{29}{35}$	13 $\frac{7}{10}$	22 $\frac{13}{22}$	31 $\frac{39}{40}$
5 $\frac{29}{30}$	14 $\frac{13}{16}$	23 $\frac{13}{15}$	32 $\frac{13}{18}$
6 $\frac{39}{56}$	15 $\frac{17}{21}$	24 $\frac{3}{4}$	33 $\frac{17}{20}$
7 $\frac{25}{42}$	16 $\frac{33}{100}$	25 $\frac{19}{20}$	34 $\frac{17}{18}$
8 $\frac{20}{21}$	17 $\frac{19}{20}$	26 $\frac{17}{24}$	35 $\frac{19}{30}$
9 $\frac{19}{42}$	18 $\frac{5}{8}$	27 $\frac{19}{20}$	36 $\frac{2}{3}$

Exercise 5h page 78

1 $\frac{2}{3}$	4 $\frac{11}{20}$	7 $\frac{5}{13}$	10 $\frac{5}{21}$
2 $\frac{1}{2}$	5 $\frac{2}{5}$	8 $\frac{3}{5}$	11 $\frac{7}{15}$
3 $\frac{5}{17}$	6 $\frac{3}{7}$	9 $\frac{5}{21}$	12 $\frac{1}{3}$

Answers

13 $\frac{18}{55}$ **16** $\frac{1}{12}$ **19** $\frac{3}{16}$ **22** $\frac{1}{4}$ **22** $17\frac{3}{7}$ **25** $15\frac{2}{5}$ **28** $17\frac{13}{32}$

14 $\frac{1}{9}$ **17** $\frac{9}{100}$ **20** $\frac{4}{15}$ **23** $\frac{1}{6}$ **23** $17\frac{3}{16}$ **26** $15\frac{4}{5}$ **29** $22\frac{2}{7}$

15 $\frac{3}{26}$ **18** $\frac{19}{56}$ **21** $\frac{1}{8}$ **24** $\frac{4}{15}$ **24** $21\frac{1}{18}$ **27** $14\frac{51}{100}$ **30** $22\frac{1}{2}$

Exercise 5i   page 79

1 $\frac{3}{8}$ **7** $\frac{3}{5}$ **13** $\frac{1}{18}$ **19** $\frac{1}{8}$

2 $\frac{5}{7}$ **8** $\frac{17}{18}$ **14** $\frac{1}{12}$ **20** $\frac{1}{3}$

3 $\frac{1}{16}$ **9** $\frac{17}{50}$ **15** $\frac{1}{5}$ **21** $\frac{19}{100}$

4 $\frac{5}{12}$ **10** $\frac{1}{2}$ **16** $\frac{1}{16}$ **22** $\frac{1}{4}$

5 $\frac{9}{50}$ **11** $\frac{3}{4}$ **17** $\frac{2}{9}$ **23** $\frac{5}{18}$

6 $\frac{5}{12}$ **12** $\frac{1}{2}$ **18** $\frac{7}{20}$ **24** $\frac{1}{30}$

Exercise 5j   page 81

1 $\frac{13}{15}, \frac{2}{15}$

2 $\frac{11}{15}, \frac{4}{15}$

3 a $\frac{1}{3}$ **b** $\frac{1}{12}$

4 a $\frac{3}{8}$ **b** $\frac{7}{8}$

5 a $\frac{11}{40}$ **b** $\frac{19}{20}$ **c** $\frac{7}{40}$

Exercise 5k   page 83

1 $2\frac{1}{4}$ **6** $3\frac{1}{2}$ **11** $13\frac{5}{8}$ **16** $12\frac{5}{6}$

2 $4\frac{3}{4}$ **7** $6\frac{3}{4}$ **12** $11\frac{6}{7}$ **17** $13\frac{2}{3}$

3 $6\frac{1}{6}$ **8** $5\frac{1}{8}$ **13** $13\frac{4}{9}$ **18** $13\frac{2}{5}$

4 $5\frac{3}{10}$ **9** $25\frac{2}{5}$ **14** $15\frac{1}{6}$ **19** $24\frac{1}{3}$

5 $9\frac{7}{9}$ **10** $10\frac{4}{11}$ **15** $7\frac{10}{11}$ **20** $4\frac{9}{10}$

Exercise 5l   page 83

1 $\frac{13}{3}$ **6** $\frac{33}{5}$ **11** $\frac{37}{5}$ **16** $\frac{73}{7}$

2 $\frac{33}{4}$ **7** $\frac{20}{7}$ **12** $\frac{22}{9}$ **17** $\frac{19}{10}$

3 $\frac{17}{10}$ **8** $\frac{25}{6}$ **13** $\frac{19}{5}$ **18** $\frac{20}{3}$

4 $\frac{98}{9}$ **9** $\frac{11}{3}$ **14** $\frac{43}{9}$ **19** $\frac{59}{8}$

5 $\frac{57}{7}$ **10** $\frac{11}{2}$ **15** $\frac{35}{4}$ **20** $\frac{101}{10}$

Exercise 5m   page 84

1 $5\frac{1}{7}$ **4** $2\frac{1}{2}$ **7** $13\frac{2}{3}$ **10** $10\frac{7}{10}$

2 $9\frac{5}{6}$ **5** $16\frac{2}{5}$ **8** $7\frac{1}{9}$ **11** $7\frac{2}{5}$

3 $4\frac{8}{11}$ **6** $7\frac{1}{4}$ **9** $8\frac{1}{6}$ **12** $6\frac{1}{2}$

Exercise 5n   page 85

1 $5\frac{3}{4}$ **8** $3\frac{3}{14}$ **15** $12\frac{1}{16}$

2 $3\frac{5}{6}$ **9** $7\frac{7}{10}$ **16** $11\frac{9}{10}$

3 $5\frac{23}{40}$ **10** $13\frac{17}{21}$ **17** $8\frac{3}{10}$

4 $9\frac{4}{9}$ **11** $10\frac{13}{16}$ **18** $18\frac{1}{2}$

5 $5\frac{29}{36}$ **12** $6\frac{1}{3}$ **19** $10\frac{1}{10}$

6 $4\frac{1}{6}$ **13** $11\frac{3}{14}$ **20** $11\frac{1}{10}$

7 $4\frac{9}{20}$ **14** $8\frac{1}{16}$ **21** $11\frac{1}{2}$

Exercise 5p   page 86

1 $1\frac{5}{8}$ **10** $3\frac{11}{35}$ **19** $3\frac{9}{35}$ **28** $2\frac{1}{2}$

2 $1\frac{13}{15}$ **11** $2\frac{2}{15}$ **20** $6\frac{2}{33}$ **29** $\frac{7}{9}$

3 $1\frac{1}{6}$ **12** $3\frac{1}{4}$ **21** $3\frac{3}{28}$ **30** $1\frac{1}{2}$

4 $\frac{3}{4}$ **13** $3\frac{3}{10}$ **22** $1\frac{5}{8}$ **31** $2\frac{5}{6}$

5 $5\frac{5}{12}$ **14** $2\frac{4}{63}$ **23** $\frac{3}{4}$ **32** $2\frac{7}{8}$

6 $1\frac{1}{2}$ **15** $3\frac{7}{24}$ **24** $1\frac{27}{35}$ **33** $3\frac{9}{10}$

7 $1\frac{5}{14}$ **16** $2\frac{25}{28}$ **25** $1\frac{3}{8}$ **34** $\frac{2}{3}$

8 $2\frac{3}{10}$ **17** $1\frac{3}{4}$ **26** $2\frac{7}{10}$ **35** $1\frac{1}{6}$

9 $1\frac{7}{10}$ **18** $3\frac{7}{20}$ **27** $3\frac{1}{2}$ **36** $2\frac{16}{21}$

Exercise 5q   page 87

1 a $1\frac{5}{21}$ **c** $\frac{35}{72}$ **e** $\frac{11}{12}$

   b $\frac{11}{24}$ **d** $2\frac{1}{6}$

2 a $2\frac{1}{4}$ **b** $3\frac{1}{5}$

3 a $\frac{3}{7}$ **b** $\frac{17}{30}$

4 a $\frac{1}{2}, \frac{3}{5}, \frac{13}{20}, \frac{7}{10}$ **c** $\frac{3}{5}, \frac{7}{10}, \frac{71}{100}, \frac{17}{20}$

   b $\frac{7}{12}, \frac{2}{3}, \frac{3}{4}, \frac{5}{6}$

5 a $<$ **b** $>$ **c** $>$

6 a $\frac{3}{11}$ **b** $\frac{7}{22}$

Exercise 5r   page 88

1 a $\frac{2}{15}$ **c** $\frac{3}{22}$ **e** $\frac{1}{2}$

   b $1\frac{7}{10}$ **d** $6\frac{7}{12}$ **f** $2\frac{13}{20}$

2 a $\frac{7}{8}$ **b** $1\frac{5}{6}$ **c** $\frac{12}{13}$

3 a $\frac{13}{100}$ **b** $\frac{233}{366}$

4 a $>$ **b** $<$ **c** $<$

5 a $\frac{3}{10}, \frac{7}{20}, \frac{3}{8}, \frac{2}{5}$ **c** $\frac{17}{32}, \frac{9}{16}, \frac{5}{8}, \frac{3}{4}$

   b $\frac{3}{10}, \frac{2}{5}, \frac{7}{15}, \frac{1}{2}$

6 a $\frac{15}{28}$ **b** $\frac{2}{7}$

Exercise 5s   page 88

1 a $\frac{43}{140}$ **c** $\frac{1}{8}$ **e** 0

   b $\frac{17}{45}$ **d** $3\frac{1}{12}$ **f** 5

2 a $1\frac{3}{8}$ **b** $2\frac{2}{5}$ **c** $\frac{5}{16}$

3 a $<$ **b** $<$

4 a $\frac{1}{2}, \frac{3}{5}, \frac{3}{4}, \frac{5}{6}$ **b** $\frac{1}{2}, \frac{5}{9}, \frac{2}{3}, \frac{5}{6}$

5 a $\frac{7}{60}$ **b** $\frac{1}{3}$ **c** $\frac{38}{79}$

6 a $\frac{17}{19}$ **b** $\frac{13}{19}$

Exercise 5t   page 89

1 a $1\frac{1}{6}$ c $\frac{1}{12}$ e $\frac{11}{12}$

    b $\frac{5}{8}$ d $2\frac{9}{20}$ f $3\frac{2}{3}$

2 a $4\frac{3}{8}$ b $\frac{1}{8}$ c $2\frac{4}{7}$

3 a $\frac{5}{24}$ b $\frac{1}{10}$ c $\frac{5}{12}$

4 a >   b <

5 a $\frac{5}{11}, \frac{1}{2}, \frac{23}{44}, \frac{13}{22}$ b $\frac{5}{9}, \frac{7}{12}, \frac{2}{3}, \frac{3}{4}$

6 a $\frac{1}{5}$ b $\frac{8}{15}$ c $\frac{1}{3}$

CHAPTER 6

Exercise 6b   page 91

1 $\frac{3}{8}$ **10** $\frac{14}{27}$ **19** $\frac{3}{4}$ **28** $\frac{1}{5}$

2 $\frac{10}{21}$ **11** $\frac{3}{20}$ **20** $\frac{6}{7}$ **29** $\frac{1}{7}$

3 $\frac{2}{15}$ **12** $\frac{3}{35}$ **21** $\frac{5}{48}$ **30** $\frac{3}{16}$

4 $\frac{7}{16}$ **13** $\frac{1}{6}$ **22** $\frac{11}{20}$ **31** $\frac{3}{20}$

5 $\frac{3}{7}$ **14** $\frac{4}{7}$ **23** $\frac{4}{11}$ **32** $\frac{2}{3}$

6 $\frac{4}{63}$ **15** $\frac{7}{18}$ **24** $\frac{4}{11}$ **33** 4

7 $\frac{6}{35}$ **16** $\frac{2}{3}$ **25** $\frac{2}{9}$ **34** $\frac{1}{18}$

8 $\frac{6}{25}$ **17** $\frac{1}{9}$ **26** $\frac{2}{31}$ **35** $\frac{3}{22}$

9 $\frac{5}{24}$ **18** $\frac{15}{28}$ **27** $\frac{2}{3}$ **36** $\frac{1}{6}$

Exercise 6c   page 92

1 $\frac{3}{5}$ **9** $16\frac{1}{3}$ **17** 10   **25** 23

2 2   **10** $\frac{17}{21}$ **18** 10   **26** 9

3 $\frac{3}{4}$ **11** 14   **19** 20   **27** 14

4 $11\frac{1}{5}$ **12** 4   **20** 60   **28** 12

5 $\frac{1}{2}$ **13** 30   **21** 7   **29** 3

6 $\frac{1}{2}$ **14** $16\frac{1}{2}$ **22** 15   **30** 8

7 $\frac{7}{8}$ **15** $7\frac{1}{2}$ **23** 5

8 2   **16** 9   **24** $6\frac{1}{3}$

Exercise 6d   page 93

1 23   **4** $37\frac{1}{2}$ **7** 36   **10** $18\frac{1}{3}$

2 30   **5** 110   **8** $8\frac{1}{2}$ **11** 14

3 $12\frac{1}{2}$ **6** $13\frac{1}{2}$ **9** 120   **12** 44

Exercise 6e   page 94

1 6   **11** 12 m   **21** 50 c

2 6   **12** 25 dollars   **22** 8 c

3 3   **13** 45 litres   **23** 30 c

4 16   **14** 33 miles   **24** 12 c

5 10   **15** 21 gallons   **25** 292 days

6 6   **16** 8 m   **26** 9 h

7 5   **17** 10 dollars   **27** 1 day

8 8   **18** 28 litres   **28** $3

9 30   **19** 15 miles   **29** 60 c

10 15   **20** 88 gallons   **30** 21 h

Exercise 6f   page 95

1 14   **11** 18   **21** $1\frac{1}{2}$

2 20   **12** 16   **22** $\frac{2}{5}$

3 21   **13** 49   **23** 1

4 15   **14** 99   **24** $2\frac{1}{3}$

5 12   **15** 39   **25** $\frac{2}{3}$

6 10   **16** 63   **26** $1\frac{1}{2}$

7 21   **17** 38   **27** $1\frac{2}{5}$

8 45   **18** $\frac{3}{4}$ **28** $2\frac{2}{3}$

9 99   **19** $1\frac{1}{5}$ **29** $3\frac{3}{8}$

10 30   **20** $\frac{1}{12}$

Exercise 6g   page 96

1 $10\frac{1}{2}$ **6** $6\frac{2}{3}$ **11** 6   **16** 6

2 $\frac{5}{6}$ **7** $\frac{9}{10}$ **12** $2\frac{2}{3}$ **17** $1\frac{3}{7}$

3 $5\frac{1}{3}$ **8** $4\frac{5}{6}$ **13** 12   **18** $3\frac{1}{3}$

4 6   **9** $1\frac{4}{5}$ **14** 6   **19** $1\frac{1}{2}$

5 $2\frac{8}{11}$ **10** 4   **15** $5\frac{3}{5}$ **20** 12

Exercise 6h   page 97

1 1   **4** $\frac{2}{3}$ **7** $5\frac{1}{10}$ **10** $\frac{9}{32}$

2 $2\frac{1}{2}$ **5** $\frac{8}{15}$ **8** $2\frac{1}{4}$ **11** $\frac{9}{20}$

3 $1\frac{2}{3}$ **6** $2\frac{2}{3}$ **9** $1\frac{1}{2}$ **12** $\frac{4}{5}$

Exercise 6i   page 98

1 $\frac{3}{5}$ **11** $\frac{2}{21}$ **21** $\frac{1}{21}$ **31** T

2 $\frac{7}{12}$ **12** $\frac{7}{10}$ **22** $1\frac{1}{4}$ **32** F

3 $\frac{1}{5}$ **13** $\frac{21}{34}$ **23** $\frac{1}{4}$ **33** T

4 $\frac{3}{14}$ **14** $1\frac{1}{2}$ **24** $\frac{1}{3}$ **34** T

5 $\frac{13}{15}$ **15** $\frac{1}{22}$ **25** $\frac{1}{9}$ **35** F

6 $\frac{5}{24}$ **16** $\frac{9}{22}$ **26** $4\frac{2}{9}$ **36** F

7 $1\frac{5}{8}$ **17** $\frac{5}{21}$ **27** $1\frac{3}{8}$ **37** T

8 $\frac{41}{42}$ **18** $\frac{5}{18}$ **28** $2\frac{7}{30}$ **38** F

9 $\frac{1}{16}$ **19** $\frac{2}{33}$ **29** $\frac{11}{6}$ **39** F

10 $\frac{1}{3}$ **20** $1\frac{2}{25}$ **30** $1\frac{8}{9}$ **40** T

Exercise 6j   page 99

1 $4\frac{3}{4}$ **11** $6\frac{1}{4}$ **21** $1\frac{1}{12}$ **31** $3\frac{1}{2}$

2 $1\frac{1}{8}$ **12** 2   **22** $\frac{7}{12}$ **32** 1

3 $\frac{3}{4}$ **13** $1\frac{7}{10}$ **23** $4\frac{1}{8}$ **33** $2\frac{1}{4}$

4 4   **14** $2\frac{2}{7}$ **24** $3\frac{1}{12}$ **34** 0

5 $2\frac{1}{14}$ **15** $2\frac{2}{5}$ **25** $\frac{7}{8}$ **35** $\frac{1}{5}$

6 $\frac{17}{18}$ **16** $\frac{17}{20}$ **26** $1\frac{3}{8}$ **36** $\frac{3}{8}$

7 22   **17** $4\frac{1}{14}$ **27** $3\frac{1}{20}$ **37** $\frac{1}{16}$

8 $\frac{13}{15}$ **18** $\frac{7}{8}$ **28** $1\frac{1}{2}$ **38** $4\frac{2}{7}$

9 3   **19** $3\frac{13}{16}$ **29** $5\frac{3}{7}$ **39** $2\frac{6}{7}$

10 $3\frac{7}{8}$ **20** $4\frac{1}{2}$ **30** $\frac{1}{2}$ **40** 2

Exercise 6k   page 100

1 30 kg   **3** 3 km   **5** 22

2 $\frac{7}{20}$ litres   **4** $58\frac{1}{2}$ min   **6** $1\frac{1}{2}$

Exercise 6l   page 101

1 a $1\frac{2}{3}$ **b** $2\frac{3}{8}$

2 6   **3** $\frac{5}{6}$ **4** $1\frac{13}{20}$ **5** $\frac{3}{5}, \frac{2}{3}, \frac{7}{10}$

6 a $19\frac{1}{3}$ **b** $1\frac{1}{2}$

7 $2\frac{1}{6}$

8 6

9 18 min

10 $3\frac{9}{10}$

11 a 27   **b** 40

12 a $2\frac{3}{5}$ **b** $3\frac{7}{8}$ **c** $5\frac{2}{5}$

13 a T   **b** T   **c** F

14 63 min   **15** $124\frac{1}{2}$ g

Exercise 6m   page 102

1 a 15   **b** $11\frac{1}{3}$

2 a $1\frac{2}{3}$ **b** $4\frac{11}{18}$

3 a <   **b** <

4 a $1\frac{1}{12}$ **b** 9

5 $\frac{1}{3}, \frac{2}{5}, \frac{7}{15}$ **6** 2

7 a $6\frac{1}{4}$ **b** $2\frac{6}{11}$

8 125 s

9 a 24   **b** 21

10 a $3\frac{1}{8}$ **b** $5\frac{4}{9}$ **c** $6\frac{1}{6}$

11 $12\frac{1}{8}$ km; $\frac{77}{97}$ **12** 6

Exercise 6n   page 102

1 a $2\frac{25}{36}$ **b** 0

2 a $\frac{1}{4}$ **b** $\frac{4}{5}$

3 25 days   **4** $\frac{17}{20}, \frac{3}{4}, \frac{7}{10}$

5 a $6\frac{1}{4}$ **b** $17\frac{11}{12}$

6 $\frac{8}{11}$ **7** $1\frac{6}{7}$ **8** $2\frac{2}{5}$

9 a $7\frac{1}{3}$ **b** $9\frac{1}{5}$ **c** $10\frac{3}{5}$

10 a, b and c   **11** 18 min   **12** $1\frac{4}{7}$ kg

CHAPTER 7

Exercise 7a   page 105

13 5.86, 4.99
14 2.29, 2.27
15 38.64, 38.46
16 6, 5.98
17 8.99, 9.45
18 3.71, 3.74
19 15.35, 15.53
20 6, 6.02

21 2.81, 6.27, 6.72
22 7.03, 7.07, 7.18, 7.41
23 0.055, 0.505, 0.55, 5.5
24 14.16, 13.55, 12.32, 10.02
25 6.555, 6.55, 6.5, 6.05
26 0.111, 0.11, 0.101, 0.011

Exercise 7b   page 106

1 $\frac{1}{5}$ **13** $\frac{73}{100}$ **25** $\frac{1}{4}$

2 $\frac{3}{50}$ **14** $\frac{81}{1000}$ **26** $\frac{9}{125}$

3 $1\frac{3}{10}$ **15** $\frac{207}{1000}$ **27** $\frac{19}{50}$

4 $\frac{7}{10000}$ **16** $\frac{29}{10000}$ **28** $\frac{61}{2000}$

5 $\frac{1}{1000}$ **17** $\frac{67}{100000}$ **29** $\frac{3}{20}$

6 $6\frac{2}{5}$ **18** $\frac{17}{100}$ **30** $\frac{1}{40}$

7 $\frac{7}{10}$ **19** $\frac{71}{1000}$ **31** $\frac{7}{20}$

8 $2\frac{1}{100}$ **20** $\frac{3001}{10000}$ **32** $\frac{1}{625}$

9 $1\frac{4}{5}$ **21** $\frac{207}{10000}$ **33** $\frac{11}{250}$

10 $1\frac{7}{10}$ **22** $\frac{63}{100}$ **34** $\frac{1}{8}$

11 $15\frac{1}{2}$ **23** $\frac{31}{1000}$ **35** $\frac{12}{25}$

12 $8\frac{3}{50}$ **24** $\frac{47}{100}$ **36** $\frac{5}{8}$

Exercise 7c   page 107

1 0.07   **5** 0.4   **9** 7.08
2 0.9   **6** 2.06   **10** 0.0006
3 1.1   **7** 0.04   **11** 4.005
4 0.002   **8** 7.8   **12** 0.0029

Exercise 7d   page 108

1 10.8   **10** 9.12   **19** 6.798
2 7.55   **11** 0.2673   **20** 27.374
3 0.039   **12** 2.102   **21** 2.38
4 3.98   **13** 0.00176   **22** 17.301
5 5.83   **14** 0.131   **23** 15.62
6 14.04   **15** 4.698   **24** 13.52
7 7.6   **16** 0.3552   **25** 16.81
8 12.24   **17** 4.6005
9 3.68   **18** 20.7

Exercise 7e   page 109

1 2.5   **13** 3.06   **25** 0.00527
2 7.8   **14** 2.94   **26** 0.05927
3 18.5   **15** 3.13   **27** 5.27
4 0.41   **16** 2.66   **28** 5.927
5 0.0321   **17** 2.4   **29** 7.24
6 16.87   **18** 7.882   **30** 729.4
7 2.241   **19** 6.118   **31** 0.72994
8 0.191   **20** 2.772   **32** 0.13
9 71.4   **21** 11.1974   **33** 57.6
10 6.65   **22** 0.000197   **34** 8.3
11 41.45   **23** 0.0067   **35** 0.149
12 6.939   **24** 0.0013   **36** 6.81

Exercise 7f   page 110

1 10.32   **8** 0.286   **15** 0.59   **22** 5.3 m
2 6.92   **9** 0.234   **16** 0.007   **23** $24.77
3 2.98   **10** 77.62   **17** 0.382   **24** $2104
4 6.6   **11** 39.88   **18** 6.64   **25** 1
5 4.4   **12** 36.52   **19** 38.82   **26** 53.2 cm
6 100.28   **13** 202.84   **20** 7.81   **27** $925
7 99.72   **14** 17.76   **21** 22.6 cm   **28** 5.9 cm

Exercise 7g   page 113

1 72 000	**5** 3278	**9** 72 810
2 82.4	**6** 430	**10** 0.000 063
3 0.24	**7** 60.2	**11** 0.703
4 460	**8** 32.06	**12** 374

Exercise 7h   page 114

1 2.772	**5** 2.7	**9** 0.0426
2 7.626	**6** 0.068	**10** 1.34
3 0.000 024	**7** 0.026	**11** 0.003 74
4 0.014	**8** 0.0158	**12** 0.0092

Exercise 7i   page 114

1 0.16	**11** 0.24	**21** 0.000 24
2 16	**12** 63	**22** 0.000 003
3 7.8	**13** 3.2	**23** 4.1
4 0.000 78	**14** 0.079	**24** 10.04
5 1420	**15** 0.078	**25** 4.2 m
6 6.8	**16** 0.24	**26** $6352
7 0.0163	**17** 11 100	**27** 0.138, 1380
8 0.002	**18** 0.000 38	**28** 0.16
9 0.14	**19** 0.0038	**29** 0.1746
10 78 000	**20** 380 000	**30** 0.0038

Exercise 7j   page 115

1 0.2	**23** 0.0057	**45** 4.55
2 1.6	**24** 0.0453	**46** 0.000 155
3 0.21	**25** 0.0019	**47** 2.35
4 2.6	**26** 0.09	**48** 0.0124
5 0.1	**27** 0.1043	**49** 0.125
6 0.19	**28** 0.000 015	**50** 0.038 75
7 0.224	**29** 0.9	**51** 0.52
8 3.8	**30** 0.0106	**52** 1.905
9 21.3	**31** 0.019	**53** 2.6
10 2.51	**32** 0.77	**54** 0.05
11 1.64	**33** 2.107	**55** 0.0025
12 0.15	**34** 0.62	**56** 0.6028
13 0.019	**35** 0.037	**57** 0.853 75
14 0.000 13	**36** 0.78	**58** 2.45
15 0.002 18	**37** 1.2	**59** 0.575
16 0.042	**38** 1.85	**60** 0.055 75
17 0.002	**39** 0.415	**61** 3.65 cm
18 0.000 06	**40** 0.15	**62** 4.075 m
19 0.81	**41** 0.72	**63** 7.15 kg
20 1.06	**42** 0.000 04	**64** 3.2 cm
21 0.308	**43** 0.8875	**65** $4.50
22 0.1092	**44** 1.75	

Exercise 7k   page 117

1 1.1	**9** 0.53	**17** 3.2
2 0.15	**10** 0.26	**18** 0.43
3 0.12	**11** 0.56	**19** 0.21
4 0.45	**12** 0.7	**20** 0.000 713
5 0.51	**13** 0.32	**21** 0.52
6 3.2	**14** 0.26	**22** 3.12
7 0.0041	**15** 0.024	**23** 0.84
8 0.036	**16** 0.000 23	**24** 0.005 68

Exercise 7l   page 118

1 0.25	**4** 0.3125	**7** 0.625	**9** 0.12
2 0.375	**5** 0.04	**8** 0.4375	**10** 0.031 25
3 0.6	**6** 2.8		

Exercise 7m   page 118

1 $\frac{1}{5}$	**5** $\frac{3}{5}$	**9** 0.9	**13** 0.03
2 $\frac{3}{10}$	**6** $\frac{7}{10}$	**10** 0.25	**14** 0.75
3 $\frac{4}{5}$	**7** $\frac{9}{10}$	**11** 0.8	**15** 0.625
4 $\frac{3}{4}$	**8** $\frac{1}{20}$	**12** 0.375	**16** 0.07

Exercise 7n   page 119

1 0.006	**5** 0.0003	**9** 0.0008
2 0.01	**6** 0.000 04	**10** 0.000 000 6
3 0.018	**7** 0.24	**11** 0.018
4 0.06	**8** 0.000 48	**12** 0.008

Exercise 7p   page 120

1 0.18	**15** 8.1	**29** 0.03
2 0.0024	**16** 0.0088	**30** 0.014 08
3 0.018	**17** 0.077	**31** 0.64
4 0.000 56	**18** 0.28	**32** 0.8
5 0.0108	**19** 0.1502	**33** 0.64
6 0.000 021	**20** 1.6	**34** 0.0008
7 0.035	**21** 1.4	**35** 6.4
8 4.8	**22** 0.000 912	**36** 0.08
9 0.0064	**23** 240	**37** 0.000 000 006 4
10 0.0018	**24** 63	**38** 800
11 0.042	**25** 0.112	**39** 0.64
12 0.72	**26** 2.048	**40** 0.008
13 0.84	**27** 22.4	**41** 0.0432
14 0.036	**28** 0.0022	**42** 12.4

Exercise 7q   page 121

1 6.72	**7** 33	**13** 64.8	**19** 0.0784
2 12.48	**8** 0.000 278 8	**14** 0.111 52	**20** 0.1054
3 0.0952	**9** 7476	**15** 0.002 592	**21** 1.722
4 1253.2	**10** 118.4	**16** 2.56	**22** 17.29
5 434	**11** 8.97	**17** 2.56	**23** 22.96
6 0.4536	**12** 198	**18** 2.56	**24** 0.031 02

Exercise 7r   page 122

1 $325	**4** 16.8 cm	**7** 3.25 m
2 4.4 cm	**5** $7471	**8** 50.4 m
3 3.8 kg	**6** 0.24	

Exercise 7s   page 123

1 $\frac{1}{50}$	**4** 2.38	**7** 0.875
2 0.009, 0.091	**5** 0.0205	**8** 20.72 cm
3 36.87	**6** 3.01	

Exercise 7t   page 123

1 $\frac{3}{10}$	**4** 0.000 62	**7** $1371.50
2 0.14	**5** 0.06	**8** $\frac{2}{5}$
3 27.32	**6** 1.5	**9** 4.75

Exercise 7u   page 124

1 $\frac{1}{125}$	**4** 85.04	**7** 0.1875
2 0.8	**5** 0.0086	**8** 4.8 cm
3 27.79	**6** 0.25	**9** 32.48

Exercise 7v   page 124

1 0.125	**4** 2.98	**7** 2.8
2 6.28	**5** $\frac{9}{100}$	**8** 2.52 m
3 0.26	**6** 280	**9** 1.224

REVIEW TEST 1   page 126

1 C	**4** A	**7** A	**10** C	**13** D
2 C	**5** D	**8** A	**11** D	
3 A	**6** B	**9** A	**12** B	

14 a $\frac{1}{9}$ **b** 10, 15   **c** $33\frac{1}{2}$

15 a 12   **b** Yes. 13 is a factor of 12 740
No. 9 is not a factor

16 a $\frac{19}{32}$ litres   **b** $\frac{1}{4}$

17 a (i) {1, 2, 3, 6}
(ii) {1, 2, 3, 4, 6, 7, 9, 12, 14, 18, 21, 36, 42}

b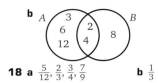

18 a $\frac{5}{12}, \frac{2}{3}, \frac{3}{4}, \frac{7}{9}$ **b** $\frac{1}{3}$

CHAPTER 8

Exercise 8a   page 130
1 a metres   **d** kilometres
  b centimetres   **e** centimetres
  c metres   **f** millimetres
3 a 3   **b** 2   **c** 4   **d** 1   **e** 6
4 (to the nearest millimetre)
  a 30   **b** 20   **c** 5   **d** 50   **e** 45
9 40 cm   **10** 900 cm

Exercise 8b   page 131
1 200   **7** 6000   **13** 150   **19** 270
2 5000   **8** 100 000   **14** 23   **20** 190 000
3 30   **9** 3000   **15** 4600   **21** 38
4 400   **10** 2 000 000 **16** 3700   **22** 9200
5 12 000   **11** 500   **17** 1900   **23** 2300
6 150   **12** 7000   **18** 3500   **24** 840

Exercise 8c   page 133
1 12 000   **7** 6000   **13** 1500   **19** 11 300
2 3000   **8** 2 000 000   **14** 2700   **20** 2500
3 5000   **9** 4000   **15** 1800   **21** 7300
4 1 000 000 **10** 2 000 000 **16** 700   **22** 300 000
5 1 000 000 **11** 3000   **17** 5 200 000 **23** 500
6 13 000   **12** 4000   **18** 600   **24** 800

Exercise 8d   page 134
1 136   **6** 3020   **11** 3500   **16** 1020
2 35   **7** 502   **12** 2008   **17** 1250
3 1050   **8** 5500   **13** 5500   **18** 3550
4 48   **9** 202   **14** 2800   **19** 2050
5 207   **10** 8009   **15** 3250   **20** 1010

Exercise 8e   page 134
1 30   **15** 3.8   **29** 1.0001
2 6   **16** 0.086   **30** 0.000 085
3 1.5   **17** 0.56   **31** 5.142
4 25   **18** 0.028   **32** 48.171
5 1.6   **19** 0.19   **33** 9.008
6 0.072   **20** 0.086   **34** 9.088
7 0.12   **21** 3.45   **35** 12.019
8 8.8   **22** 8.4   **36** 4.111
9 1.25   **23** 11.002   **37** 1.056
10 2.85   **24** 2.042   **38** 5.003
11 1.5   **25** 4.4   **39** 0.2505
12 3.68   **26** 5.03   **40** 0.850 55
13 1.5   **27** 7.005
14 5.02   **28** 4.005

Exercise 8f   page 136
1 5.86   **10** 2456   **19** 2580   **28** 73.6
2 1.035   **11** 109   **20** 2362   **29** 2642
3 3001.36   **12** 1358   **21** 2.22   **30** 19 850
4 3051   **13** 3250   **22** 1606.4   **31** 35 420
5 5.647   **14** 5115   **23** 1089.6   **32** 910
6 4.65   **15** 15 100   **24** 5972   **33** 448.2
7 440   **16** 2550   **25** 748   **34** 5
8 55   **17** 1046.68   **26** 0.922
9 1820   **18** 308.73   **27** 1150

Exercise 8g   page 138
1 13 540   **5** 32   **9** 22.77
2 45 792   **6** 10.6   **10** 16 240
3 13.563   **7** 15 366
4 12.55   **8** 24.448

Exercise 8h   page 139
1 9.72 m   **5** 1080 mm   **9** 33.2 cm
2 1840 g   **6** 4 kg   **10** 5.3 kg
3 748 kg   **7** 2.2 kg
4 4.11 g   **8** 15 m

Exercise 8i   page 141
1 a September   **b** Thursday
  c 17th   **d** 13th
2 13   **3** 4
4 a Dennis   **b** Johanne   **c** 2023
5 a 17   **b** P. Baldrick   **c** 2011
6 a 3 hours 10 min   **b** 18 days 18 hours
7 a 308 sec   **b** 210 min
8 a $\frac{1}{3}$ **b** $\frac{1}{100}$
9 a 10.25   **b** 11.10   **c** 45 min
10 a 21 min   **b** 1734
11 a 1 hour 45 min   **b** 8 hour 40 min
  c 2 hour 10 min
12 6.05 p.m.
13 2 hours 40 min
14 a 20 min   **b** 8.05 p.m.
15 a 6 hours 30 min   **b** 8 hours 29 min
  c 8 hours   **d** 27 hours 3 min
16 1715
17 a 2 hour 22 min and 2 hour 17 min
  b Astleton and Morgan's Hollow; shortest journey time

Exercise 8j   page 145
1 a 12 °C   **b** 27 °F
  c the one in part a; it is above the freezing point of water, the other is below
2 a 105 °F   **b** 40 °C   **c** 20 °C   **d** 68 °F
3 a 68 °F   **b** 40 °F   **c** 27 °C   **d** 2 °C
4 38.3 °C

Exercise 8k   page 146
1 a grams   **b** seconds   **c** metres
  d kilometres   **e** metres
  f degrees Celsius (or Fahrenheit)
  g tonnes   **h** centimetres
  i millimetres   **j** weeks
  k seconds   **l** kilograms
2 a 98 cm   **b** 980 mm
3 2.23 km   **6** 3.6 m
4 3056 m, 3050 m   **7** 95 t 660 kg; 121 t 960 kg
5 9.192 kg   **8** 76.9 kg, 72 kg
9 78.8 km, 1.2 km   **10** 113 °F
11 1905   **12 a** 1.06 p.m.   **b** 24 min

Exercise 8l   page 148
1 4000 m   **5** 3000 cm   **9** 0.065 kg
2 0.03 kg   **6** 1.25 km   **10** 4.29 kg
3 350 cm   **7** 1.5 m
4 0.25 kg   **8** 28 mm

Exercise 8m   page 148
1 2.36 m   **4** 0.5 g   **7** 2.35 kg
2 20 mm   **5** 4.25 km   **8** 2000 mg
3 5000 g   **6** 3600 kg   **9** 2.6 m

Exercise 8n   page 149

1 5780 kg
2 354 cm
3 0.35 t
4 0.0155 cm
5 1.56 t
6 7.80 m
7 90 min
8 2.05 km
9 8.598 t

Exercise 8p   page 149

1 4.2 cm
2 0.35 kg
3 1520 g
4 0.283 km
5 3.6 cm
6 470 mm
7 0.36 m
8 3 days
9 350

CHAPTER 9

Exercise 9a   page 151

1 68 in
2 14 ft
3 1809 yd
4 35 in
5 100 in
6 4320 yd
7 17 ft
8 123 in
9 28 ft
10 118 in
11 3 ft
12 2 ft 5 in
13 7 ft 2 in
14 3 yd
15 4 yd 1 ft
16 1 mile 240 yd
17 6 ft 3 in
18 33 yd 1 ft
19 10 ft
20 17 miles 80 yd

Exercise 9b   page 152

1 38 oz
2 28 oz
3 67 oz
4 64 cwt
5 162 lb
6 1 lb 8 oz
7 1 lb 2 oz
8 2 lb 4 oz
9 1 ton 10 cwt
10 1 cwt 8 lb

Exercise 9c   page 154

1 6 lb
2 6 ft
3 2 kg
4 3 m
5 3 lb
6 15 ft
7 7 lb
8 $2\frac{1}{2}$ m
9 8 oz
10 1 lb
11 16 km
12 32 km
13 24 km
14 160 km
15 120 km
16 64 km
17 11 lb
18 2 m
19 2 m
20 4 kg
21 37.5 miles
22 8 oz
23 15 cm
24 4 in

25 a 25 mm   **b** 15 mm
26 15 cm   **27** in the market

CHAPTER 10

Exercise 10a   page 158

2 a 2   **b** 1   **c** 4   **d** 3
3 a ray   **b** line   **c** line segment
4 a triangle (irregular)
   b pentagon (regular)
   c quadrilateral (irregular)
   d square (regular)
   e hexagon (irregular)

Exercise 10b   page 159

1 $\frac{3}{4}$
2 $\frac{1}{2}$
3 $\frac{1}{4}$
4 $\frac{1}{2}$
5 $\frac{1}{4}$
6 $\frac{1}{2}$
7 $\frac{1}{2}$
8 $\frac{1}{2}$
9 1
10 $\frac{1}{4}$
11 $\frac{1}{3}$
12 $\frac{1}{3}$
13 $\frac{3}{4}$
14 $\frac{3}{4}$
15 $\frac{2}{3}$
16 6
17 9
18 9
19 3
20 6
21 6
22 4
23 8
24 9
25 12

Exercise 10c   page 160

1 N
2 W
3 N
4 E No
5 N
6 $\frac{3}{4}$
7 $\frac{3}{4}$
8 $\frac{1}{2}$

Exercise 10d   page 161

1 1
2 2
3 3
4 1
5 4
6 2
7 3
8 4
9 2
10 1
11 1
12 3
13 4

Exercise 10e   page 162

1 obtuse
2 acute
3 reflex
4 acute
5 obtuse
6 reflex
7 acute
8 acute
9 obtuse
10 acute
11 reflex
12 obtuse
13 obtuse
14 obtuse
15 acute

Exercise 10f   page 163

1 180°
2 90°
3 270°
4 180°
5 90°
6 270°
7 180°
8 270°
9 90°
10 120°
11 270°
12 270°
13 180°
14 90°
15 180°
16 30°
17 45°
18 120°
19 60°
20 45°
21 30°
22 120°
23 30°
24 60°
25 120°
26 210°
27 180°
28 300°
29 330°
30 150°
31 210°
32 300°
33 210°
34 150°
35 210°

Exercise 10g   page 165

1 34°
2 60°
3 75°
4 137°
5 150°
6 20°
7 115°
8 54°
9 80°
10 11°
11 325°
12 332°
13 250°
14 218°
15 345°
16 330°
17 240°
18 345°
19 282°
20 213°
21 145°

Exercise 10h   page 168

1 30°
2 60°
3 90°
4 120°
5 150°
6 180°
7 3
8 2
9 4
10 12
11 5
12 9
13 1
14 10
15 2
16 6
17 3
18 7
19 6
20 8
21 1
22 12
35 60°
36 140°
37 350°
38 260°
39 25°
40 300°
41 45°
42 5°
43 25°
44 80°
45 160°
46 105°

Exercise 10j   page 171

1 180°   **2** 180°

Exercise 10k   page 172

1 120°
2 155°
3 10°
4 100°
5 20°
6 130°
7 80°
8 15°
9 135°
10 140°
11 90°
12 50°
13 AB, CD; EF, GH
14 a LM, NP   **b** 145°
15 a ST and MN or PQ   **b** 90°
16 105°, 180°
17 40°, 180°   **19** 140°   **21** 90°, 180°
18 320°   **20** 115°, 180°

Exercise 10l   page 174

1 240°
3 20°
4 a PQ and RS or TU and VW
   b PQ and TU or VW, or RS and TU or VW
5 140°
6 140°

Exercise 10m   page 175
1 240°   **3** 354°   **5** 50°
2 W   **4** 140°, 40°   **6** 30°

CHAPTER 11

Exercise 11a   page 178
1, 3, 4 and **6**

Exercise 11b   page 181
1 2   **3** 0   **5** 2
2 1   **4** 1   **6** 2

Exercise 11c   page 183
1 6   **2** 6   **3** 0   **4** 3

Exercise 11d   page 185
2, 3, 5 and **6**
9 In Exercise 11c, numbers 1, 2, 3, 4 and 7 have
rotational symmetry.

CHAPTER 12

Exercise 12a   page 188
1 $\frac{1}{5}$ **13** $\frac{5}{8}$ **25** $\frac{41}{50}$ **37** 3.5
2 $\frac{9}{20}$ **14** $\frac{5}{4}$ **26** $\frac{7}{8}$ **38** 0.485
3 $\frac{1}{4}$ **15** $\frac{7}{10}$ **27** $\frac{1}{16}$ **39** 0.92
4 $\frac{18}{25}$ **16** $\frac{3}{4}$ **28** $\frac{3}{2}$ **40** 0.65
5 $\frac{1}{3}$ **17** $\frac{12}{25}$ **29** 0.47   **41** 1.2
6 $\frac{1}{8}$ **18** $\frac{69}{100}$ **30** 0.12   **42** 2.31
7 $\frac{1}{40}$ **19** $\frac{3}{8}$ **31** 0.055   **43** 0.85625
8 $\frac{1}{2}$ **20** $\frac{4}{75}$ **32** 1.45   **44** 0.08
9 $\frac{13}{20}$ **21** $\frac{7}{40}$ **33** 0.5875   **45** 0.03
10 $\frac{14}{25}$ **22** $\frac{19}{20}$ **34** 0.58   **46** 1.8
11 $\frac{37}{100}$ **23** $\frac{3}{20}$ **35** 0.3   **47** 0.052
12 $\frac{2}{3}$ **24** $\frac{2}{25}$ **36** 0.623   **48** 0.54125

Exercise 12b   page 189
1 50%   **11** 75%   **21** 50%   **31** 25%
2 70%   **12** 45%   **22** 22%   **32** 74%
3 65%   **13** 140%   **23** 83%   **33** 125%
4 25%   **14** $62\frac{1}{2}$%   **24** 172%   **34** 341%
5 52.5%   **15** 94%   **25** 62.5%   **35** 7.5%
6 25%   **16** 60%   **26** 90%   **36** 36%
7 15%   **17** 35%   **27** 4%   **37** 16%
8 16%   **18** 124%   **28** 55%   **38** 139%
9 37.5%   **19** $87\frac{1}{2}$%   **29** 264%   **39** 635%
10 46%   **20** 160%   **30** 84.5%   **40** 18.25%

Exercise 12c   page 190
1 a $\frac{3}{10}$ **b** $\frac{17}{20}$ **c** $\frac{17}{40}$ **d** $\frac{21}{400}$
2 a 0.44   **b** 0.68   **c** 1.7   **d** 0.165
3 a 40%   **b** 85%   **c** $12\frac{1}{2}$%   **d** 106.25%
4 a 20%   **b** 62%   **c** $84\frac{1}{2}$%   **d** 178%

	Fraction	Percentage	Decimal
	$\frac{3}{4}$	75%	0.75
5	$\frac{4}{5}$	80%	0.8
6	$\frac{3}{5}$	60%	0.6
7	$\frac{7}{10}$	70%	0.7
8	$\frac{11}{20}$	55%	0.55
9	$\frac{11}{25}$	44%	0.44
10	$\frac{8}{25}$	32%	0.32

Exercise 12d   page 190
1 52%   **7** 43%   **13** 1400
2 13%   **8** 68%   **14 a** 2%
3 36%   **9** 20%   **b** 10%
4 92%   **10** 38%   **c** 66%
5 88%   **11** 3%   **d** 22%
6 12%   **12** 252

Exercise 12e   page 192
1 25%   **13** 200%   **25** 72%
2 60%   **14** $62\frac{1}{2}$%   **26** 42%
3 $33\frac{1}{3}$%   **15** 10%   **27** 40%
4 $33\frac{1}{3}$%   **16** $66\frac{2}{3}$%   **28** 65%
5 75%   **17** 25%   **29** $23\frac{1}{3}$%
6 60%   **18** $37\frac{1}{2}$%   **30** $66\frac{2}{3}$%
7 15%   **19** 20%   **31** $2\frac{1}{2}$%
8 25%   **20** 40%   **32** 36%
9 10%   **21** 60%   **33** $33\frac{1}{3}$%
10 20%   **22** 25%   **34** 4%
11 30%   **23** 72%   **35** $13\frac{1}{3}$%
12 50%   **24** $333\frac{1}{3}$%   **36** $2\frac{1}{2}$%

Exercise 12f   page 193
1 48   **13** 333   **25** 320 m²
2 96 g   **14** 198 kg   **26** 45 km
3 55.5 cm   **15** 1.44 m   **27** 5 km
4 286 km   **16** $1.50   **28** 149 cm²
5 16 c   **17** 0.34 km   **29** 14 c
6 3.08 kg   **18** 1.6 litres **30** $53.63
7 252   **19** $75   **31** 48 c
8 989 g   **20** 198 m   **32** 6 g
9 4.73 m   **21** 90 g   **33** 2.1 m
10 206.4 cm²   **22** 2.94 mm   **34** $10
11 2.52 m   **23** 18 cm   **35** 2 kg
12 14.4 m²   **24** 9 m²   **36** 14 mm

Exercise 12g   page 194
1 40%   **6** 75%
2 70%   **7** 75%
3 20%   **8** $66\frac{2}{3}$%
4 20%   **9** 65%
5 30%   **10** 1960
11 a $46\frac{2}{3}$%   **b** $53\frac{1}{3}$%
12 a 52   **b** 28
13 a 12   **b** 18
14 a 3   **b** 147
15 5760   **16** 78   **17** $62.40   **18** 112

Exercise 12h   page 196

1 a $\frac{2}{5}$ **b** $\frac{27}{50}$ **c** $\frac{11}{40}$

2 a 60%   **b** 78%   **c** $12\frac{1}{2}$%

3 8%   **4** $12\frac{1}{2}$%   **5** 54 ml   **6** 97%

Exercise 12i   page 196

1 a $\frac{9}{25}$ **b** 0.36

2 a 62.5%   **b** 125%   **c** 250%

3 $12\frac{1}{2}$%   **4** 289 m²   **5** $44 000

Exercise 12j   page 197

1 a $12\frac{1}{2}$%   **b** $37\frac{1}{2}$%   **c** 50%

2 a 40%   **b** 27.9%   **c** 162.5%

3 a $\frac{1}{8}$ **b** 0.125

4 90 c   **5** 54

CHAPTER 13

Exercise 13a   page 199

1 700 c   **10** 504 p   **19** 11.09 c
2 600 p   **11** £1.26   **20** £6.08
3 800 c   **12** $3.50   **21** £3.20
4 1300 c   **13** £1.90   **22** $5.05
5 735 c   **14** €3.50   **23** $1920
6 4381 c   **15** $43.07   **24** €6
7 1103 c   **16** £2.28   **25** $2.10
8 615 p   **17** €3.47
9 210 p   **18** $5.80

Exercise 13b   page 200

1 $114   **7** $4.50   **13** $680
2 $900   **8** $85.50   **14** $25 000
3 $7644   **9** $5   **15** $535
4 $870   **10** $720   **16** $48
5 $21 440   **11** $96   **17** $600
6 $1692   **12** 25
18 a $27 618   **b** $4603

Exercise 13c   page 202

1 Single peppers; $29 against $37
2 Pack of 6; $58 per can against $60 per can
3 The larger bag; $1.25 per clip against $1.4 per clip
4 1 kg pack; $0.37 per gram against $0.47 per gram
5 Yes; 1 km costs $250 against $260
6 Prepacks; $650 per kg against $704 per kg
7 a $40   **b** $36.25   **c** 200 g jar
  d no, it costs $37 for 25 g which is dearer than the 200 g jar
8 950 g jar; $16.66 per 50 g against $15.78 per 50 g

Exercise 13d   page 204

1 $550
2 $1495
3 $600 000
4 a $136 profit   **b** $26 775 profit
  c $110 loss   **d** $1408 loss
5 $11 500
6 $144 000
7 a $2880   **b** yes, $7130
8 a $5400   **b** $7565   **c** $2165   **d** $3500

Exercise 13e   page 205

1 45%   **3** 80%   **5** 16%
2 120%   **4** 30%   **6** $22\frac{1}{2}$%

7 12%
8 a $2025   **b** 50%
9 85%
10 a $43200   **b** 400%
11 a $480   **b** 36%
12 a $391000   **b** 15%
13 $37\frac{1}{2}$%   **14** 80%
15 $33\frac{1}{3}$%   **16** $12\frac{1}{2}$%
17 a $2560   **b** 55%
18 a $6000   **b** 12%
19 a $4550   **b** 65%
20 a $4000   **b** $66\frac{2}{3}$%
21 a a loss of $5000   **b** 10% loss
22 $12\frac{1}{2}$% gain   **23** 60%

Exercise 13f   page 206

1 $52 200
2 a 1500 c   **b** $463
3 $50 040
4 Shop B. In Shop A 1 g costs $1.30 and in Shop B 1 g costs $1.20
5 55%   **6** 20%

Exercise 13g   page 207

1 $24 890
2 a 1550 p   **b** £6.50
3 $190
4 The jar in Bestway. In Mates 1 g costs 75 c and in Bestway 1 g costs 70c.
5 a $2880   **b** 40%
6 a $3600   **b** 60%

CHAPTER 14

Exercise 14a   page 210

1 A(2, 2)   B(5, 2)   C(7, 6)   D(4, 5)   E(7, 0)
  F(9, 4)   G(0, 8)   H(5, 8)
4 square
5 isosceles triangle
6 rectangle
7 square
8 isosceles triangle
9 5   **12** 1   **15** 5   **18** 6
10 7   **13** 14   **16** 4   **19** 5
11 0   **14** 0   **17** 1   **20** 0
21 (9, 12), (9, 9), (13, 6)
22 (3, 11), (3, 7), (7, 7); 4
23 (1, 1), (6, 1), (8, 4), (3, 4); 5, 5
24 (13, 3); 4
25 (2, 5)   **27** (4, 1)   **29** (3, 7)
26 (7, 1)   **28** (5, 4)   **30** (2, 3)

Exercise 14b   page 214

1 a 8, 8, 8, 8   **b** DC, yes   **c** 90°
2 a AB and DC, BC and AD
  b AB and DC, BC and AD
  c 90°
3 a all equal
  b AB and DC, BC and AD
  c $\hat{A} = \hat{C}$, $\hat{B} = \hat{D}$
4 a AB and DC, BC and AD
  b AB and DC, BC and AD
  c $\hat{A} = \hat{C}$, $\hat{B} = \hat{D}$
5 a none   **b** AB and DC   **c** none

Answers

Exercise 14c   page 215

1 parallelogram
2 rectangle
3 trapezium
4 square
5 trapezium
6 rhombus
7 square
8 rectangle
9 parallelogram
10 rhombus

REVIEW TEST 2   page 217

1 C **4** A **7** C **10** C **13** C
2 C **5** D **8** C **11** C
3 B **6** B **9** B **12** C

14 $(9, 3)$, $\left(7\frac{1}{5}, 2\frac{2}{5}\right)$
15 6 cm
16 a $a = 40°$, $b = 180°$
17 950 g jar
18 a 0.281 25 **b** 3 kg for \$16.50

CHAPTER 15

Exercise 15a   page 221

1 a 180 m **b** 6
2 a 5 m **b** 4
3 600 cm **4** 5 m
5 a a square **b** 140 cm
6 10 cm
7 9 cm **8** 13 cm
9 8 cm **10** 8 cm
11 11 cm
12 a 2 cm **b** 2 cm **c** 5 m **d** 9 mm
e 21 cm **f** 3.8 cm **g** 6.3 m
13 a 16 cm, 15.7 cm **d** 15 cm, 14.4 cm
b 30 cm, 29.5 cm **e** 14 cm, 16.8 cm
c 145 mm, 143 mm **f** 12 cm, 12.22 cm
14 a 23 cm **b** 20.04 cm **c** 1080 mm **d** 58.8 m

Exercise 15b   page 224

1 11 **9** 45
2 16 **10** 45
3 11 **11** 45
4 20 **12** 40
5 26 **13** 37
6 20 **14** 76
7 21 **15** 62
8 a A **b** B **16** 26

Exercise 15c   page 228

1 4 cm² **8** 1.44 cm² **14** 280 cm²
2 64 cm² **9** $\frac{1}{4}$ km² **15** 3.96 mm²
3 100 cm² **10** $\frac{9}{16}$ m² **16** 1470 km²
4 25 cm² **11** 30 cm² **17** 2.85 m²
5 2.25 cm² **12** 48 cm² **18** 30.24 cm²
6 6.25 cm² **13** 27 m² **19** 22 800 cm²
7 0.49 m² **20** 36 000 mm²

Exercise 15d   page 228

1 120 cm² **5** 52 m² **9** 43 m²
2 36 m² **6** 87 cm² **10** 228 cm²
3 149 m² **7** 544 mm²
4 208 mm² **8** 90 cm²

Exercise 15e   page 230

1 a 24 cm **b** 28 cm²
2 a 24 cm **b** 24 cm²
3 a 48 mm **b** 80 mm²
4 a 32 m **b** 15 m²
5 a 272 cm **b** 1664 cm²

6 184 cm² **9** 432 cm²
7 91 cm² **10** 4.84 m²
8 198 cm²

Exercise 15f   page 232

1 4 **3** 6 **5** 45
2 9 **4** 6 **6** 500

Exercise 15g   page 234

1 a 30 000 **b** 120 000 **c** 75 000
d 820 000 **e** 85 000
2 a 1400 **b** 300 **c** 750
d 2600 **e** 3250
3 a 560 **b** 56 000
4 a 4 **b** 25 **c** 0.5
d 0.25 **e** 7.34
5 a 0.55 **b** 14 **c** 0.076
d 1.86 **e** 2970
6 a 7.5 **b** 0.43 **c** 0.05
d 0.245 **e** 176

Exercise 15h   page 235

1 50 000 cm² **6** 15 000 cm²
2 1800 mm² **7** 37 500 cm²
3 175 000 cm² **8** 180 mm²
4 14 000 cm² **9** 120 000 m²
5 8 m² **10** 22 500 m²

Exercise 15i   page 235

1 a 370 m **b** 8250 m²
2 a 340 m **b** 7000 m²
3 a 380 m **b** 8400 m²
4 a 76 m **b** 312 m²
5 5 m² **7** \$900 **9** 100
6 1200 **8** 9000 cm² **10** 96

CHAPTER 16

Exercise 16a   page 239

1 a sphere **b** cube **c** cone
d hemisphere, cylinder **e** hemisphere **f** cube, cuboid
g cuboid **h** cylinder **i** cuboid
j cylinder **k** sphere **l** cylinder
m cuboid **n** cuboid **p** cylinder
q cylinder, cone **r** cuboid **s** cylinder
2 cuboid, cylinder
4 square pyramid
5 a cuboid **b** cuboid, triangular prism
c cube, square pyramid **d** cuboid
e cuboid, cylinder **f** cylinder, cone
g cylinder, hemisphere

Exercise 16b   page 241

2 a i 2 **ii** 2 **iii** 4 cm by 3 cm
3 a 6
4 b IJ **c** G and K
5

6 b There are 36 arrangements altogether.
c 11 will make a cube.

CHAPTER 17

Exercise 17a   page 247

1

2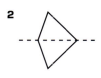

3

4

5

6

7

8

9

10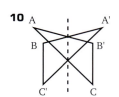

11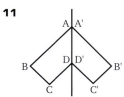

12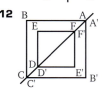

13

14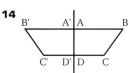

15

16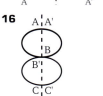

17

18

19 Q9: A and A', Q11: A A'; D D', Q12: A A'; F, F'; D, D'; C, C', Q13: B, B', Q14: A, A' and D, D', Q15: C, C', Q16: A, A'; B, B'; C, C, Q17: A, A'; C, C'.
They all lie on the axis of symmetry.

20 Equal distances; perpendicular lines.

21 Equal distances; perpendicular lines.

Exercise 17b   page 250

1 The *x*-axis.

2 The line through $y = 1$ parallel to the *x*-axis.

3 The line through $x = 2$ parallel to the *y*-axis.

4 The line through the origin at 45° to the axes.

5 a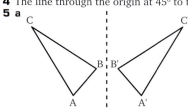

b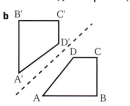

c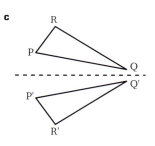

Exercise 17d   page 252

1 b, d

2 a, c

3 1

4 b, c, d

5 a, b

6 b, d

CHAPTER 18

Exercise 18a   page 260

1 $x - 3 = 4, 7$

2 $x + 1 = 3, 2$

3 $3 + x = 9, 6$

4 $x - 5 = 2, 7$

5 $2x = 8, 4$

6 $7x = 14, 2$

7 $3x = 15, 5$

8 $6x = 24, 4$

Exercise 18b   page 262

1 8	**7** 6	**13** 10	**19** 5
2 9	**8** 6	**14** 3	**20** 12
3 2	**9** 5	**15** 8	**21** 12
4 7	**10** 7	**16** 10	**22** 3
5 4	**11** 3	**17** 9	**23** 2
6 5	**12** 1	**18** 12	**24** 9

Answers

Exercise 18c page 263

1 2	**10** 4	**19** 5	**28** 23
2 9	**11** 4	**20** 11	**29** 4
3 3	**12** 8	**21** 16	**30** 7
4 13	**13** 12	**22** 12	**31** 9
5 3	**14** 10	**23** 10	**32** 9
6 3	**15** 11	**24** 9	**33** 4
7 7	**16** 10	**25** 17	**34** 4
8 0	**17** 6	**26** 5	**35** 4
9 1	**18** 11	**27** 16	**36** 2

Exercise 18d page 264

1 2	**7** $\frac{1}{3}$	**13** 6	**19** 9
2 3	**8** 3	**14** 1	**20** 2
3 $2\frac{1}{2}$	**9** $1\frac{2}{5}$	**15** $\frac{1}{6}$	**21** $\frac{3}{4}$
4 3	**10** 20	**16** 2	**22** $1\frac{1}{5}$
5 4	**11** 2	**17** $1\frac{4}{5}$	**23** 5
6 $2\frac{1}{4}$	**12** $\frac{1}{2}$	**18** $3\frac{1}{2}$	**24** $\frac{1}{7}$

Exercise 18e page 264

1 4	**8** 16	**15** 5	**22** 0
2 12	**9** $5\frac{1}{2}$	**16** $\frac{2}{7}$	**23** 5
3 2	**10** 13	**17** $2\frac{2}{3}$	**24** 20
4 1	**11** 8	**18** 7	**25** 30
5 $1\frac{1}{5}$	**12** 16	**19** 2	**26** 30
6 3	**13** 6	**20** $1\frac{2}{3}$	**27** $\frac{1}{5}$
7 8	**14** $3\frac{1}{3}$	**21** 11	

Exercise 18f page 265

1 4	**10** 7	**19** 2	**28** 3
2 3	**11** 5	**20** 2	**29** $2\frac{1}{5}$
3 2	**12** 3	**21** $1\frac{2}{3}$	**30** $\frac{1}{3}$
4 6	**13** 5	**22** $\frac{1}{2}$	**31** 6
5 3	**14** 2	**23** 4	**32** $\frac{1}{4}$
6 0	**15** 3	**24** 0	**33** 5
7 6	**16** 3	**25** 2	**34** $\frac{6}{7}$
8 5	**17** 0	**26** $3\frac{1}{3}$	
9 $2\frac{2}{3}$	**18** $1\frac{4}{5}$	**27** $2\frac{3}{7}$	

Exercise 18g page 266

1 $4x-8=20$, 7	**6** $2x+6=24$, 9
2 $6x-12=30$, 7	**7** $2x+6=20$, 7
3 $3x+6=21$, 5	**8** $2x+10=24$, 7
4 $x+8=10$, 2	**9** $3x-9=18$, 9
5 $3x+7=28$, 7	**10** $2x+9=31$, 11 cm

Exercise 18h page 268

1 $10x$	**3** $2x$	**5** $8y$	**7** 1
2 $4x$	**4** 2	**6** 7	**8** 0

Exercise 18i page 269

1 $7x+7$	**7** $4x+2y$	**13** $10x+8y$
2 $5x+5$	**8** $4x+8y$	**14** $11x+y$
3 $4x+1$	**9** $8x+3$	**15** $15x$
4 $5c+10a$	**10** $8x+8$	**16** $4x+y+8z$
5 $8x+2y$	**11** $3x+6$	**17** $9x+y+3$
6 $8x+8y$	**12** $19x+3y$	**18** 11

Exercise 18j page 270

1 4	**3** 3	**5** $\frac{7}{3}$	**7** 6
2 1	**4** 5	**6** $-\frac{3}{4}$	**8** 5

Exercise 18k page 271

1 1	**9** $\frac{1}{2}$	**17** 7
2 1	**10** 2	**18** 2
3 4	**11** $1\frac{2}{3}$	**19** 5
4 $1\frac{6}{7}$	**12** 2	**20** 1
5 6	**13** $1\frac{1}{5}$	**21** $\frac{1}{2}$
6 2	**14** 3	**22** 10
7 $4\frac{1}{2}$	**15** 2	
8 2	**16** 1	

Exercise 18l page 272

1 $\frac{2}{3}$	**4** 2
2 $x+4=10$; 6	**5** $9x-y$
3 2	**6** $1\frac{1}{3}$

Exercise 18m page 272

1 2	**4** 4
2 $7c$	**5** $4\frac{1}{3}$
3 $1\frac{1}{2}$	**6** $6a+1$

Exercise 18n page 272

1 $5\frac{1}{2}$	**4** 0
2 0	**5** $3b$
3 2	**6** $14-x=8+x$; 3

Exercise 18p page 272

1 4	**4** 2
2 $-x$	**5** $2a+5c+5d$
3 $-\frac{2}{5}$	**6** $\frac{2}{5}$

CHAPTER 19

Exercise 19a page 275

1 $2x+2$	**7** $5a+5b$
2 $9x+6$	**8** $16x+12$
3 $5x+30$	**9** $18+12x$
4 $12x+12$	**10** $5x+5$
5 $8+10x$	**11** $14+7x$
6 $12+10a$	**12** $24+16x$

Exercise 19b page 275

1 $6x+4$	**5** $5x+23$
2 $10x+18$	**6** $5x+5$
3 $3x+7$	**7** $3x+5$
4 $4x+17$	**8** $9x+8$

Exercise 19c page 276

1 2	**6** 2	**11** $\frac{1}{3}$
2 0	**7** 3	**12** 2
3 $1\frac{3}{8}$	**8** $2\frac{1}{2}$	**13** 2
4 3	**9** $\frac{2}{7}$	**14** $4\frac{2}{5}$
5 1	**10** 3	

Exercise 19d   page 277

1 11
2 6
3 9 cm
4 $21
5 16
6 $20
7 4
8 80°
9 6

Exercise 19e   page 280

1 $2l+2w$
2 $2l+d$
3 $3l$
4 $5l$
5 $2l+s+d$
6 $W=x+y$
7 $P=2l-2b$
8 $T=N+M$
9 $T=N-L$
10 $A=l\times!$
11 $N=10n$
12 $C=nx$
13 $L=l-d$
14 $P=6l$
15 $A=2l\times l$
16 $N=S-T$
17 $W=T+S$
18 $S=N-L-R$
19 $r=p-q$ or $r=q-p$
20 $W=Kn$
21 $d=b-a$

22 $q=20x$
23 $L=\frac{ny}{1000}$
24 $A=100lb$
25 $T=t+\frac{s}{60}$

Exercise 19f   page 282

1 10
2 100
3 30
4 2
5 20
6 200
7 24
8 15
9 25
10 $7\frac{1}{2}$
11 $1\frac{5}{5}$
12 $\frac{1}{2}$
13 $3\frac{1}{3}$
14 7
15 $1\frac{3}{4}$
16 19
17 0

Exercise 19g   page 283

1 $C=50n$, $600
2 $L=50n$, $500
3 $P=2a+2b$, 70 cm
4 $P=6x$, 6 cm
5 $P=L-Nr$, 5 m
6 $W=Na+p$, 45
7 $A=2lw+2lh+2hw$, 6200 cm²

Exercise 19h   page 284

1 $x=5$
2 $4x+13$
3 13
4 $x=2$
5 3
6 11
7 $P=3l+2h+g$

Exercise 19i   page 285

1 $x=\frac{1}{2}$
2 $10x+15$
3 12
4 $x=0$
5 $x=1$
6 30
7 $P=6a$

Exercise 19j   page 285

1 $x=\frac{1}{2}$
2 $6+x+12=4x$; $x=6$
3 $2x+7$
4 $x=1$
5 a $N=2000+20m+10t$
  b $11 000
6 $x+x+2+8=18$; $4
7 $3x+10$
8 We get $3=0$ which cannot be true

CHAPTER 20

Exercise 20a   page 289

1 7, 14, 17, 22, 12
2 4, 22, 18, 17, 7, 2, 1, 1
3 1, 2, 10, 15, 16, 20, 10, 6, 2

Exercise 20c   page 292

1 a lowest = 1987; highest: 1992
  b 1992
  c 1992, 1996
  d 1993, 1994
2 a Antigua   **b** 1987
  c 1989, 1990, 1991
3 a 2006–7   **d** 2008–9
  b defence   **e** 2008–9
  c 2007–8

Exercise 20d   page 293

1 b 1995   **c** 1986
2 b Dominica   **c** Barbados

Exercise 20e   page 295

1 96°, 132°, 60°, 42°, 30°
2 128°, 152°, 48°, 24°, 8°
3 303°, 3°, 30°, 24°
4 84°, 204°, 48°, 24°
5 144°, 48°, 80°, 88°
6 60°, 84°, 48°, 36°, 84°, 48°
7 96°, 120°, 36°, 72°, 36°
8 108°, 180°, 40°, 18°, 14°
9 72°, 13.5°, 85.5°, 94.5°, 54°, 40.5°
10 223°, 40°, 54°, 36°, 7°
11 61°, 82°, 82°, 21°, 10°, 103°

Exercise 20f   page 297

1 a business and professional
  b (i) $\frac{1}{12}$ (ii) $\frac{7}{36}$
2 a heating   **b** a little less
3 a (i) $\frac{1}{6}$ (ii) $\frac{1}{4}$
  b 20–39 and 40–59; under 10 and 60–79

Exercise 20g   page 298

1 a 10, 14, 10, 22   **b** danger
  c very effective
2 a Spanish   **b** 18, 15, 11, 12, 16: total 72   **c** Acceptable
3 a consumption is rising each year
  b impression is given by the volume in the bottle which goes up more quickly than the height of the bottle.

CHAPTER 21

Exercise 21a   page 303

1 2, {H, T}
2 3, {R, B, Y}
3 10, (1, 2, 3, 4, 5, 6, 7, 8, 9, 10}
4 6, {R, Y, B, Brown, Black, G}
5 3, {chewing gum, boiled sweets, bar of chocolate}
6 4, {1 c, 10 c, 20 c, 50 c}
7 13, {A, 2, 3, 4, 5, 6, 7, 8, 9, 10, J, Q, K}
8 5, {a, e, i, o, u}
9 5, {2, 3, 5, 7, 11}
10 10, {2, 4, 6, 8, 10, 12, 14, 16, 18, 20}

Exercise 21b   page 304

1 $\frac{1}{4}$
2 $\frac{1}{10}$
3 $\frac{1}{5}$
4 $\frac{1}{6}$
5 $\frac{1}{7}$
6 $\frac{1}{200}$
7 $\frac{1}{52}$
8 $\frac{1}{7}$
9 $\frac{1}{15}$

Exercise 21c   page 306

1 5   **2** 3   **3** 26   **4** 2   **5** 10

6 a $\frac{1}{2}$ **b** $\frac{1}{2}$ **c** $\frac{2}{5}$ **d** $\frac{3}{10}$

7 a $\frac{1}{13}$ **b** $\frac{1}{2}$ **c** $\frac{1}{4}$ **d** $\frac{4}{13}$

8 a $\frac{2}{9}$ **b** $\frac{1}{9}$ **c** $\frac{1}{3}$ **d** $\frac{2}{9}$

9 a $\frac{1}{2}$ **b** $\frac{1}{3}$ **c** $\frac{1}{3}$

10 $\frac{2}{15}$

11 a $\frac{3}{5}$ **b** $\frac{1}{5}$ **c** $\frac{2}{5}$

12 $\frac{1}{40}$

13 a $\frac{18}{37}$ **b** $\frac{18}{37}$ **c** $\frac{9}{37}$

14 $\frac{21}{26}$

15 $\frac{4}{45}$

16 a $\frac{5}{12}$ **b** $\frac{1}{3}$ **c** $\frac{3}{4}$

Exercise 21d   page 308

1 0, impossible
2 0.3, unlikely to be this heavy
3 1, almost certain
4 0.001, possible but unlikely
5 0, most unlikely!
6 0, impossible
7 1, certain
8 0, virtually impossible
9 1, it must be
10 0, almost impossible
11 Likely: you will watch TV this week, you will get maths homework this week.
Unlikely: you will be a millionaire; it will snow in Jamaica on Christmas day.

Exercise 21e   page 309

1 $\frac{3}{5}$ **3** $\frac{21}{26}$ **5** $\frac{7}{10}$ **7** $\frac{24}{25}$ **9** $\frac{39}{40}$

2 $\frac{12}{13}$ **4** $\frac{5}{6}$ **6** $\frac{5}{8}$ **8** $\frac{2}{3}$ **10** $\frac{10}{13}$

11 a $\frac{1}{10}$ **b** $\frac{3}{10}$ **c** $\frac{2}{5}$ **d** $\frac{7}{10}$

12 a $\frac{1}{13}$ **b** $\frac{1}{4}$ **c** $\frac{3}{4}$ **d** $\frac{11}{13}$

13 a $\frac{15}{22}$ **b** $\frac{7}{22}$ **c** $\frac{1}{22}$ **d** $\frac{3}{11}$

14 a $\frac{2}{5}$ **b** $\frac{19}{30}$ **c** $\frac{7}{30}$ **d** 0

Exercise 21f   page 311

1 about 50
2 10
3 20
4 a 10   **b** 20
5 20
6 a 15   **b** 45   **c** 30
7 20
8 10
9 10
10 5
11 25
12 a $\frac{1}{13}$ **b** 3
13 a $\frac{1}{2}$ **b** 30
   c no, possible but not certain.
   d Probably, you'd expect about 50.
14 a $1000   **b** 5   **c** $500
   d lose, spending greater than likely winnings.

15 a 2
   b $5000
   c Spends £5000 more than winnings.
16 a 2
   b $2000
   c $2560
   d no, she spends $560 more than she is likely to win.
17 yes. likely loss $4800, likely win $5200

REVIEW TEST 3   page 316

1 B   **3** C   **5** C   **7** C   **9** D   **11** D   **13** C
2 C   **4** D   **6** B   **8** C   **10** A   **12** B

14 a $n+2$
   b $3m+6$
   c $x=7$

15 a

Weight	Number
41–45	1
46–50	3
51–55	11
56–60	9
61–65	4
66–70	3

b

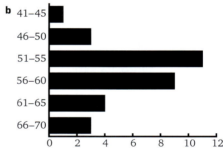

16 40 cm²

17

Mark	Frequency
1	1
2	2
3	1
4	3
5	5
6	0
7	3
8	2
9	2
10	3

REVIEW TEST 4   page 320

1 B   **11** C   **21** D
2 C   **12** A   **22** B
3 A   **13** A   **23** D
4 D   **14** A   **24** A
5 D   **15** D   **25** B
6 B   **16** B   **26** B
7 D   **17** B   **27** B
8 C   **18** D   **28** B
9 C   **19** D   **29** C
10 C   **20** B   **30** B

Index

A

accuracy of measurement 137–8
acute angles 162, 165, 176
addition 4, 6, 12
    continuous 4
    of decimals 107–8, 124
    of fractions 75, 76–7, 79, 80, 81
    of metric quantities 136, 139
    of mixed numbers 84–5
age distribution 297
algebra 259
alumina production 293
a.m. 141
angles 160–1, 162
    acute 162, 165, 176
    drawing 170
    measuring 164–5
    obtuse 162, 170, 176
    reflex 162, 166, 167, 176
    right 161, 163, 171
    on a straight line 171, 172, 173
answers 324–42
approximations 14–15, 23, 131, 137–8, 311
Arabic numerals 259
Archimedes 237
area 224
    compound figures 228, 230, 231, 236
    irregular shapes 224, 225, 236
    rectangles 227, 235, 236
    squares 227, 232, 236
    units of 226–7, 233–5, 236
Aryabhata 276–7
axes of symmetry 177, 178–9, 181,
    183, 185
    see also mirror lines

B

Babylonians 164
balance 261, 263
bar charts 290, 291, 292–3, 299, 301
barter 129
bearings, compass 160, 169
best buys 201–2, 207
bias 307, 311, 312–13
billions 56
birth rates 294
bisector, perpendicular 251, 258
Bless My Dear Aunt Sally 30, 98, 282
boiling point, water 144

brackets 29–30, 31, 98
    curly (braces) 41
    multiplying out 275, 276
Brahmagupta 65
bricks 238

C

calculator, using 20, 22
calendars 140, 141, 273
cancelling fractions 74, 75, 91
capacity 240, 245
cards (playing) 304, 305, 309
carrying numbers 6, 25
Cartesian geometry 216, 274
Celsius 144, 145
centimetres 129, 130, 131, 135
    cubic 240
    square 227, 233, 234
cents 199
centuries 140
certainty 308, 315
chain of numbers 11
chance 304, 313
checking answers 4, 265, 273
chord 157
circles, properties 157
circumference 157
class projects 300
clock, 12-hour 141
clock, 24-hour 140, 143, 144
clock hands 159, 160, 161, 174
closed loops 157
coinage 129
coins, flipping 311
collecting like terms 268, 269, 270–1
column of numbers, adding 4, 6
common denominator 77, 106
common factors 60, 74, 75, 91
    highest (hcf) 60, 61, 64
common multiple, lowest (lcm) 61, 62, 64
comparing prices 201–2
compass directions 160, 169
composite numbers 58
compound figures
    area 228, 230, 231, 236
    perimeter 221, 230
cones 238
congruency 178–9, 186
continuous addition 4

conversion
    imperial units, length 151
    metric/imperial 153–4, 155
coordinates 208–9
cost, multiple 200, 207
cost price 203, 207
cost per unit 200, 201–2, 207
counting on 9
cross-number puzzle 13, 29
cubes (solids) 238, 240–1
cubic centimetres 240
cubic metres 240
cubic millimetres 240
cuboids 238
    nets 241, 242
currencies, writing 199
curved lines 157
cylinders 238

D
date format 142
days 140
decades 140
decimal places 119–20, 125
decimal point 105, 107, 125
    invisible 108
decimals 104–5
    addition of 107–8, 125
    division of 115, 116, 117, 125
    to fractions 106, 124
    multiplication 119–20, 121, 125
    ordering 105
    to percentages 189, 197
    recurring 119
    standard 118, 125
    subtraction of 109, 125
degrees (angular) 163, 164
denominator 66, 70, 75, 76, 82
    common 77, 106
Descartes, René 216, 274
diagonals 286, 322
diameter 157
dice 238, 241, 303, 304, 311
difference (subtracting) 10, 13
digits 1
dividing by fractions 95, 97, 103
divisibility tests 58–9
division
    by 10, 100, 1000 etc 26, 113–14
    of decimals 115, 116, 125
    by fractions 95, 97, 103
    long division 27, 117
    with whole numbers 24–5, 96, 97, 103

Dodgson, Charles 40, 192
dollars 199
drawing angles 170

E
elements (members) of sets 41, 43
equally likely 304
equations 259–60, 273
    with brackets 275–6
    with letter terms on both sides 269–70
    with like terms 270–1
    solving 261–2, 263, 273, 277–8
        in two operations 265, 266
equilateral triangles 158, 184, 244
equivalent fractions 69–70, 71, 72, 77
estimating 14–15, 23, 131, 137–8, 311
Eureka can 240
euros 199
even numbers 59
events 303, 304, 308, 309, 311
expectation 311
experiments 303, 305–6
expressions 98, 268, 269, 273
eye colour 294, 295

F
factors 57, 60, 63
    common 60, 74, 75, 89, 91
        highest (HCF) 60, 61, 64
Fahrenheit 144, 145
fair 307, 311, 312–13
fathoms 152
feet (measurement) 150, 155
formulae 279–80, 283, 286
    substituting into 282, 283
fractions 66, 84
    adding 75, 76–7, 79, 80, 81
    cancelling 74, 75, 91
    to decimals 107, 117–18, 119, 125
    dividing 95, 97, 103
    equivalent 69–70, 71, 72, 77
    multiplying 90, 97, 103
    ordering 72, 73
    to percentages 189, 197
    proper and improper 82, 83
    of quantities 68, 89, 94
    simplifying 74, 75, 89, 91
    standard 118, 125
    subtracting 78, 79, 80, 81
    whole numbers as 93
freezing point, water 144
frequency (of a group) 290, 301
frequency columns 288

frequency tables 287–8, 301
fuel use 297
furlongs 151, 152

G
gallons 240
games 36, 170, 303
grams (g) 132, 136, 149

H
handshakes 80
hemispheres 238
hexagons 157, 158, 224
highest common factor (hcf) 60, 61, 64
hindu problem solving 276–7
hours 140
hundredths 104, 105
hundredweights 152, 155

I
images 247, 248, 250, 258
imperial units 129, 150–1, 152
    metric conversion 153–4, 155
impossibility 308, 315
improper fractions 82, 83, 85
inches 150, 155
intersecting lines 171
invariant lines 249
invariant points 249
inversion 277
inverting fractions 95
isolating x 265
isosceles triangles 183

K
kilograms (kg) 132, 136, 149
kilometres 129, 130, 131, 135
    square 227, 233

L
leap years 140
length, units of
    imperial 150, 151
    metric 129–30, 133, 134
like terms 268, 269, 273
    collecting 268, 269, 270–1, 275
likely, equally 304
line segments 157, 158, 176
line symmetry 177, 178–9, 180, 181, 183, 185
lines 157, 176
    curved 157
    intersecting 171
    invariant 249

Ox and Oy 209
    parallel 171, 173, 176, 214
    perpendicular 171, 173, 176, 250
    of symmetry 177, 178–9, 181, 183
litres 240
long division 27, 117
long multiplication 21–3, 24
    of decimals 119, 121
loops, closed 157
loss 203, 204
lowest common multiple (LCM) 61, 62, 64, 72
lowest terms 74, 106

M
magic squares 5, 33, 112
magic triangles 34
map game 170
mass, units of 132–3, 152
mathematical reflection 248
mathematicians
    Asian 65, 259, 276–7
    British 40, 49, 192
    European 41, 216, 237, 244, 274
measurements, accuracy 137–8
measuring angles 164–5
measuring length 130, 138
measuring temperature 144–5
members (elements) of sets 41, 43
    common 49, 50, 51, 53, 54, 55
metres 129, 130, 131, 135, 220
    cubic 240
    square 226, 227, 233, 234
metric conversion 131–3, 134, 135–6
    from imperial 153–4, 155
'metric foot' 147
metric system 129–30
metric units of length 129–30, 133, 134
metric units of mass 132–3, 138
midday and midnight 141
miles 150, 155
    nautical 152
millennia 140
milligrams (mg) 132
millilitres (ml) 240
millimetres (mm) 129, 131, 135
    cubic 240
    square 227, 233, 234
millions 56
minutes 140
mirror lines 247, 248, 249–50, 251, 258
    see also axes of symmetry
mirrors 246–7
misleading information/charts 291

mixed operations 28, 98
mixed numbers 82, 83, 84–5, 86
    dividing with 96, 97, 103
    multiplying 92, 93, 97, 103
Moebius strip 244
Mohammed ibn Musa al-Khowarizmi 259
money 129, 198, 199
months 140
multiple cost 200, 207
multiples 43, 57, 62
    lowest common (lcm) 61, 62, 64, 72
    of x 263–4
multiplication
    by 10, 100, 1000 etc 20–1, 112–13
    of decimals 119–20, 121, 125
    with fractions 90, 97, 103
    long 21–3, 24, 119, 121
    of metric units 138
    of mixed numbers 92, 93, 97, 103
    of whole numbers 18–19, 20

N
nautical miles 152
nets 240–1, 242, 243
Nightingale, Florence 287
noon 141
number pairs (coordinates) 209
number patterns 5–6, 34, 35, 36
numbers, ordering 2
numbers, square 36
numbers, unknown (x) 259, 260, 261, 268
numerals 1, 16, 259
numerator 66, 70, 74, 76, 82

O
object 247, 248, 250, 258
obtuse angles 162, 170, 176
octagons 224
odds 303, 304
of (multiply) 90, 91, 94, 103
operations 79
    mixed 28, 98
    order of 28, 29–30, 31, 98
ordering decimals 105
ordering fractions 72, 73
ordering numbers 2
origin 209
ounces 152, 153, 155
outcome of experiments 303

P
palindromes 8–9
parallel lines 171, 173, 176, 214

parallelograms 214, 215
patterns, number 5–6, 34, 35, 36
patterns, strip 254–5
patterns, textile 256–8
pence 199
pentagons 116, 157, 224
percentage profit and loss 204–5
percentages 190, 193, 195, 197
    to decimals 188, 197
    to fractions 187–8, 197
    of a quantity 192, 194, 197
perfect numbers 63
perimeter 110, 221, 236
    compound figures 221, 230
    quadrilaterals 223
    rectangles 267, 277
    squares 221, 266, 280
perpendicular bisector 251, 258
perpendicular lines 171, 173,
    176, 250
pets owned 300
pictographs 298, 299, 300, 301
pie charts 294, 295–6, 301
    interpreting 297
pints 240
place value 1–2, 104, 105
plane shapes 157
plane surfaces 157, 171, 238
planets' weights 284
playing cards 304, 305, 309
plotting points 208–9
p.m. 141
points 157, 208–9, 251
    of intersection 171
    invariant 249
    plotting 208–9
polygons 157, 158, 224
pounds (mass) 152, 153, 155
pounds (money) 199
price, comparing 201–2
price, cost 203, 207
price, selling 203, 207
prime numbers 42, 58, 64
    relatively prime 60
prisms 238, 243
probability 303, 304, 309, 315
product (multiply) 61, 91
profit and loss 203, 207
    percentage 204, 207
projects, class 300
proper fractions 82
protractors 164, 165, 170
puzzles 13, 26, 82, 87, 92

pyramids 35, 238, 243
Pyramids 239

Q
quadrilaterals 157, 158, 184, 213
    perimeter 223
    special 215
quotient 61

R
radius 157
random 304, 306
rays 157, 176
rectangles 158, 214, 215
    area 227, 235, 236, 279
    perimeter 267, 277
rectangular numbers 36
rectilinear solids 238
recurring decimals 119
reflection 246–7, 248, 250, 251, 254, 256
    mathematical 248
reflex angles 162, 166, 167, 176
relatively prime numbers 60
remainder 24, 25, 26, 27, 84
revolutions 159, 163, 164, 176
rhombuses 184, 214, 215, 322
right angles 161, 163, 171
Roman numbers 16
rotational symmetry (s-symmetry) 185, 186
roulette 307
rounding numbers 15, 23
rule for sequences 34

S
safety distance, cars 293
scratch cards 313
seconds 140
selling price 203, 207
sequence of numbers 5–6, 34, 35, 36
set notation 41
sets 40, 41, 45, 49, 55
    empty (null) 46, 47, 55
    intersection of 49, 50, 51, 53, 54, 55
    of outcomes 303
    subsets 47, 48, 49, 52, 55
    symbols 40, 41, 43, 46, 47, 50, 53, 55
    union of two 50–1, 52, 55
    universal 47–8, 49–50
shapes, properties 157
significant figures 188
simplifying expressions 268, 269
simplifying fractions 74, 75, 89, 91
simulation 314

solids 238, 239
spheres 238
square centimetres 227, 233, 234
square kilometres 227, 233
square metres 226, 227, 233, 234
square millimetres 227, 233, 234
square numbers 36
squares (shapes) 158, 210, 215
    area 227, 232, 236
    perimeter 221, 266
s-symmetry (rotational symmetry) 185, 186
standard decimals and fractions 118, 124
straight lines 157
strip patterns 254–5
subsets 47, 48, 49, 52, 55
substituting into formulae 282, 283
subtraction 9
    of decimals 109, 125
    of fractions 78, 79, 80, 81
    of mixed numbers 86
    of whole numbers 9, 10, 12, 13
sugar price 292
symbols
    mathematical 15, 72, 136, 153, 187, 270, 308–9
    sets 40–1, 43, 46–7, 50, 53, 55
symmetry, axes of 177, 178–9, 181, 183, 185
    see also mirror lines
symmetry, rotational (s-symmetry) 185, 186

T
tally columns and marks 288
tangrams 258
tans 258
temperature 144–5
    conversion 145, 147
ten-thousandths 105
tenths 104, 105
terms 74, 268, 269–70
    like 268, 269, 273
        collecting 268, 269, 270–1, 275, 277
    lowest 74, 106
    unlike 268, 273
tetrahedrons 244
textile patterns 256–8
thermometers 145
thousandths 105
three-dimensional shapes 238, 239
Tic-Tac-Toe 302
time 140–1
    'reflected' 180
timetable, bus 144
tithes 155

tonnes (t) 132, 136, 149
tons 152, 155
tourist numbers 292
trapeziums 215
triangles 157, 243
    equilateral 158, 184, 244
    isosceles 183
triangular numbers 37

U
unit cost 200, 201–2, 207
units of length, imperial 150, 151
units of length, metric 129–30, 133, 134
units of mass 132–3, 152
units of time, changing 140
unknown number (x) 259, 260, 261, 268
unlike terms 268, 273

V
value for money 201–2, 207
variables 259, 286
Venn diagrams 49–50, 51, 52, 54, 321
Venn, John 49
vertices 157, 165, 248
volume 240, 245

W
weeks 140
whole numbers
    adding 4, 6, 12
    dividing 24–5, 96
        into decimals 115, 116, 125
    to fractions 93
    multiplication 18–19, 20, 103
    subtraction 9, 10, 12, 13

X
x (unknown number) 259, 260, 261, 268
    isolating 265
    multiples of 263–4
x-axis 209
x-coordinate 209, 210, 216

Y
yards 150, 155
y-axis 209
y-coordinate 209, 210, 216
years and leap years 140

Z
zeros (noughts) 109, 114, 115, 116, 121, 128